PREVENTIVE MAINTENANCE FOR THE PILOT AND AIRCRAFT OWNER

by

Dale Crane

A practical handbook of the airplane, its engine and systems, written by a professional maintenance technician, to help you get the most efficient and economical operation from your airplane.

TABLE OF CONTENTS

A good working relationship with a competent A&P can make pilot-performed maintenance an enjoyable part of owning and flying an airplane.

- Figure 1-

PREFACE

Pilot ground school training is by nature strong on meteorology, navigation, traffic procedures, and the Federal Aviation Regulations pertaining to flying airplanes. The lack of time, however, precludes more than a cursory treatment of the airplane itself.

The Aviation Maintenance Foundation, Inc. (AMFI), the nation's largest non-profit organization for professional aviation maintenance personnel, has been concerned about the lack of practical information available to the pilot. Information regarding the operation and care of airplane and engine has not been available, heretofore, in the concise form needed by you, the man who flies the plane.

PREVENTIVE MAINTENANCE FOR THE PILOT AND AIRCRAFT OWNER is not intended to turn you into a mechanic, nor is it intended to bore you with unimportant details of airplane design and construction; nor, finally, does it use your valuable time repeating basic principles with which you, a pilot, are already familiar.

PREVENTIVE MAINTENANCE FOR THE PILOT AND AIRCRAFT OWNER does deal with the following topics:

I. Legal Aspects of Aviation Maintenance:
What really constitutes airworthiness, and how a pilot-owner can determine that his airplane meets all of the pertinent airworthiness requirements.

II. Considerations in Buying an Airplane:
What you should look for when buying a used airplane, so you will not become the proud owner of more hidden liability than you bargained for.

III. An Introduction to Aircraft Systems:
The different types of construction used for small general aviation aircraft are set forth, along with its landing gear, control systems, fuel, hydraulic, electric, instrument, and avionic systems. These are discussed, not from the operational point of view - the way they are used - but from what goes on inside.

IV. An Introduction to Powerplant Systems:
A practical presentation of the mechanical system, cooling system, lubrication, ignition, fuel, and induction systems, and the propeller. This helps you understand what is going on under the cowling so that you may more intelligently assess performance deterioration.

V. What the I.A. Looks for on an Annual Inspection:
Since the safety and longevity of an airplane is directly related to its maintenance care, each certificated airplane must have a thorough inspection every twelve months. A typical general aviation airplane is taken through an annual inspection, explaining those things the inspector looks for in order to keep the airplane airworthy.

VI. Preventive Maintenance - or What the Pilot Can Do with the Blessings of the FAA:
F.A.R. 43 lists twenty-five things a pilot can do to increase the service life of his airplane and lower the cost of maintenance. The "governmentese" of this regulation is translated into practical language and these items explained so they can be performed in an efficient and economical way.

VII. Ground Rules for Pilot-Performed Maintenance:
The FAA allows a person not holding an A&P mechanic license to do only a few things to his airplane. If you do some of your own maintenance, there are certain ground rules which, if observed, will make your relationship with the fixed base operator and the A&P who does your routine maintenance much more profitable and enjoyable. A pilot trying to save a few dollars by doing maintenance outside his legal or practical capacity seriously jeopardizes his flying enjoyment. Good rapport with a knowledgeable A&P technician can make your pilot-performed maintenance both safe and rewarding. Tools required for preventive maintenance and that all-important matter of parts are discussed in this section.

VIII. Fine Points of Aircraft Engine Operation:
You proved you could operate an aircraft engine before you were awarded your pilot's license. To operate it even more efficiently, here are some fine points on engine operation in such matters as proper use of carburetor heat and the mixture control. Magneto checking as well as engine cooling are also discussed.

Basic safety has been designed into your airplane, and the manufacturer has gone to great lengths to build it to meet high standards of safety and utility. The fixed base operator and your A&P mechanic have specialized equipment and professional training to maintain your airplane in its original state. But, ultimately, the safety and utility of the airplane lie with you, the pilot. AMFI has as its goal the promotion of safety and efficiency in aviation through the propagation of maintenance information. The aim of this book is to help you get the most utilization and enjoyment from your airplane.

INTRODUCTION

The question may be reasonably asked, "Why should an organization of professional aviation mechanics write a book on aircraft maintenance for the pilot?" The answer is quite simple. AMFI has as its main goal that of promoting safety in aviation through the propagation of maintenance information. This certainly involves the pilot-owner, the one most concerned with the maintenance of his plane. Secondly, AMFI has its eye on the general well-being of aviation itself: The more efficiently the pilot-owner utilizes his airplane, the more flying he will do; the more flying, the more airplanes; and therefore aviation grows.

This book is not written for the professional pilot whose airplane is on a continuing maintenance schedule, but for the pilot-owner who must make his own decisions as to what maintenance should be done. A knowledge of the mechanics of the airplane, its powerplant, and systems will help you, the pilot, evaluate any malfunction and will help make your decisions more accurate. And from the standpoint of economics, there are certain maintenance functions which the pilot-owner may perform on the airplane he operates, thus resulting in quite a saving on the maintenance bill. These functions are clearly discussed in the latter portion of this book.

Pilot ground school training must, by its very nature, concentrate on such matters as weather, navigation, and FARs. The vast amount of this information the pilot **must** receive from ground school precludes more than a cursory introduction to the airplane and its engine. Yet, the airplanes you fly today have such complex systems that in-depth knowledge is required for their efficient operation.

This AMFI manual is not intended to make a mechanic of the pilot, or to bore you with immaterial details of the complexities of airplane design or construction. It is, rather, to help you, the busy pilot-owner, better understand your airplane; to help you more easily and accurately analyze performance deterioration or malfunction so you can help your A&P restore the airplane to its proper functioning in quick time.

Airplane inspection and maintenance is quite involved because of the lightweight, high-performance airplanes in use. Complex electrical and electronic equipment add to the problems in which possibility of failure must be minimized. To further complicate aviation maintenance, regulations of the Federal Aviation Administration must be adhered to in detail.

This book is arranged to present aircraft maintenance for the pilot in such a way that it will help make your flying safer and more enjoyable. At intervals throughout the text, questions are asked for you to check your progress. Be sure to answer each of these questions before going on to the next section, as much of this material builds on what has been previously covered. Correct answers will be found at the end of the text for you to verify your answer. Some of the words used in the text may be new to you, so you will find them defined in the Glossary. Rather than the definitions found in conventional dictionaries, these definitions may be more in line with the language of the A&P. The first time words in the Glossary appear in the text, they are marked with an asterisk (*).

For an over-all review of your comprehension of this material, there is included a twenty question, multiple choice examination.

Careful inspection by a competent A&P does more to extend the life of an airplane than any other aspect of aviation maintenance.

- Figure 2 -

SECTION I: LEGAL ASPECTS OF AVIATION MAINTENANCE

Since your first instructional flight, you have been aware of the presence of the Federal Aviation Administration* - the FAA. In flight, you are under the cognizance of the Operations branch. In dealing with the mechanical aspects of aviation, you must comply with the requirements of the maintenance section of the FAA.

A. TYPE CERTIFICATION REQUIREMENTS:

When your airplane was designed and plans for it were drawn, the manufacturer had to prove to the FAA that the prototype* airplane built from these plans could meet all of the requirements for safety which have been established for the issuance of the Airworthiness Certificate.* Quality control in the factory assures us that each production airplane will meet these requirements as did the prototype. When the airplane leaves the factory and enters the general aviation fleet, the responsibility for it continuing to meet these safety requirements rests on the shoulders of the owner or operator.

To help the owner maintain his airplane in a condition which will meet its requirements, the FAA has established a program of inspections and records. Specifications, drawings, and manuals are available to the A&P technician* for use as maintenance information. All repairs and adjustments must be made according to this data.

B. APPROVED DATA:

1. Type Certificate Data Sheets:*

Each I.A.*, an A&P technician holding an Inspection Authorization,* is required to have available a set of Type Certificate Data Sheets covering the airplane he is inspecting. These data sheets are published by the Government Printing Office and outline the actual specifications of each model of airplane which has been issued an Approved Type Certificate.* This information includes:

a. A list of all approved engines.
b. Engine power ratings, including such limitations as take-off power*, RPM*, or manifold pressure*.
c. Airspeed limits.
d. Center of gravity* ranges.
e. Maximum weight.
f. Number of seats and their location.
g. Fuel capacity and the location of each tank.
h. Oil capacity and location of the tank, or tanks.
i. Control surface movements.
j. Location of the datum* used for weight and balance purposes.
k. Method used to level the airplane.
l. Any additional limitations or information the FAA may consider important.

There are separate Type Certificate Data Sheets for engines and for propellers. For engines, such data as approved carburetors and carburetor settings, approved spark plugs, and allowable temperatures and pressures are given. Propeller data sheets include such items as allowable hub and blade combinations and blade repair limits.

AIRWORTHINESS DIRECTIVE COMPLIANCE RECORD N 21366

Aircraft, Engine, Propeller, Rotor, or Appliance Make CESSNA Model 150 Ser.No. 150-

AD Number	Subject	Date and Hours at Compliance	Method of Compliance	One-Time	Recurring	Next Comp. Due Date/ Hours	Authorized Signature and Number
67-3-1	MUFFLER	1-31-67 985	INSPECTION	X			D. W. Evans 691916 A&P
67-3-1	GLOVE BOX	12-30-67 1215	REMOVED BOX	X			D. W. Evans 691916 A&P
71-22-2	NOSE GEAR FORK	12-12-71 1425	REPLACED FORK	X			D. W. Evans 691916 A&P
72-3-3	FLAP JACK SCREW	4-12-72 1532	INSPECTION		X	1630	D. W. Evans 691916 A&P

The FAA requires that a record be kept of all applicable Airworthiness Directives, the date, and method of their compliance.

- Figure 3-

2. Airworthiness Directives:*

If any serious defect shows up after an airplane or engine is in service, rather than cancel the Type Certificate approval, the FAA issues an Airworthiness Directive (A.D.) against the defective component. This is a mandatory directive and is only issued after it has been determined that this condition definitely compromises safety of flight. A.D.s are mailed or telegraphed to the registered owner of each affected aircraft, and all I.A.s are required to maintain an up-to-date file of all A.D.s. The directive specifies the required fix for the condition and the time limit in which the fix must be accomplished. After complying with the A.D. note, the A&P must record the method of accomplishment and the date in the aircraft records.

3. Alteration Records:

a. Major alterations:

If any major alteration or change has been made in an aircraft or engine, it no longer complies with the information in the Type Certificate Data Sheets. In order for the I.A. to certify the airplane as airworthy, there must be a record of approval for the alteration. A "Major Repair or Alteration" form, better known as Form 337,* must be filled out each time such a change is made. The original copy of this form must be kept in the aircraft records, and a copy is sent to the FAA in Oklahoma City, Oklahoma. When an FAA agent approves the alteration as described on the Form 337, this particular airplane is therefore changed from that described in the data sheet, and it **must** now include the alteration described on the form.

b. Minor alterations:

Required for each aircraft is an equipment list which includes all components approved for installation in that airplane. As an example, all radio equipment for which an approved installation has been made is included in this list. If an A&P installs any of this equipment in the airplane using the approved installation methods, it is classified as a minor alteration, and a Form 337 need not be filled out. The equipment must be marked on the list as having been installed, and the weight and balance record changed to reflect the new empty weight* and empty weight center of gravity.*

4. Manufacturer's publications:

Type Certificate Data Sheets, Airworthiness Directives, and alteration records all specify conditions the airplane must meet to maintain its airworthy status, but detailed **specific** requirements are those issued by the manufacturer for each component. Clearance, tolerances, pressures, and dimensions are usually applicable to one specific model of component and must be clearly specified and available to the A&P technician. Specific information published by the manufacturer takes precedent over general information in determining the airworthiness of a part.

QUESTIONS:

1. Who is responsible for seeing that an airplane is maintained in an airworthy condition?
2. Where would you find a listing of all the spark plugs approved for your engine?
3. How does an airplane owner know if there is an Airworthiness Directive against his airplane?
4. What record must be made of a major alteration to an airplane?
5. What record must be made of compliance with an A.D. note?

C. INSPECTION REQUIREMENTS:

The pilot is required to conduct an adequate preflight inspection of his aircraft before each flight, and the manufacturer specifies twenty-five and fifty hour inspections to increase the longevity of the airplane; but the FAA requires only the annual inspection for non-commercially operated aircraft. One hundred hour inspections are also required for aircraft operated for hire.

1. Annual Inspection:

Once each 12 calendar months* all certificated airplanes must have a thorough inspection. The airframe* is opened up and carefully checked for indication of corrosion,* cracks, or other damage. The landing gear,* brakes, tires, and wheels are critically inspected, and all functioning systems are checked for proper timing, adjustment, and operation. This inspection must be accomplished by an A&P holding an Inspection Authorization, who, after determining that the airplane meets all of its requirements for certification, will return the airplane to service for another twelve months. The scope of the annual inspection will be covered in detail later in this book.

2. One Hundred Hour Inspection:

If the airplane is flown for hire, the same inspection is required each 100 hours of operation. The only difference between the 100 hour inspection and the annual is the fact that an A&P without an Inspection Authorization can return an airplane to service after a 100 hour inspection. Regardless of how many 100 hour inspections an airplane has in a year, however, it must still have an annual each year. Since the inspections are the same, any 100 hour inspection performed by an I.A. may be considered an annual and written up as such.

3. Progressive Inspection:

Though not common for personal aircraft, progressive inspections may be used when an operator does not want to keep an airplane down for the length of time necessary to allow a complete inspection at one time. An I.A. plans an inspection schedule in which the airplane will be given a complete inspection within a 12 month period. This schedule is approved by the FAA, and after the airplane has been given a complete annual inspection, the progressive schedule starts. The inspection continues throughout the inspection period, and at its completion, the annual is considered done, and ready for the cycle to repeat.

4. Continuing Inspection Program:

Used at present, primarily on large aircraft operating under F.A.R. 91, subpart D, is a maintenance schedule which may possibly spread to cover private aircraft. It is an inspection program written for each specific operational condition and for each aircraft. This program replaces both the 100 hour inspection and the annual but is much more comprehensive and, since it is tailored to the individual operation, is more meaningful.

QUESTIONS:

6. Who can return an airplane to service after an annual inspection?
7. What is the difference between an annual inspection and a 100 hour inspection?

D. RECORD KEEPING REQUIREMENTS:

1. Maintenance records:

One has only to look at a sampling of logbook* entries to realize that the record keeping system which has been used for years has degenerated to an almost meaningless requirement. So in August of 1972, the maintenance record keeping requirements were changed and the need for "permanent" bound records was dropped.

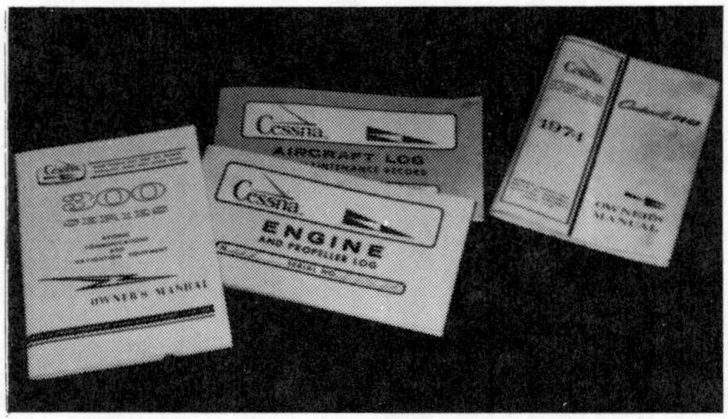

Bound logbooks are no longer required, but are still a convenient way to keep the required maintenance records.

- Figure 4 -

Although some airplane owners will still continue to keep records in bound log books, it is no longer **required.** What is required is that certain records be kept for one year, or until the work is repeated or superseded.

a. Shop records of 100 hour inspections.
b. Shop records of any major repairs, with the reference that a Form 337 having this date covers this repair. The 337 need be kept for only one year.

Certain records, however, must be kept with the airplane and transferred at the time of sale. These are:

a. Total time in service of the airframe.
b. Current status of life-limited parts. These life limits, if any apply, are spelled out in the Type Certificate Data Sheets.
c. Time since overhaul of all parts which must be overhauled on a time basis. These times are spelled out in FAA documents and are not necessarily the manufacturer's time between overhaul recommendations.
d. Identification of the current inspection status of the aircraft.
e. Current status of applicable Airworthiness Directives and their method of compliance.
f. A list of all of the current major alterations that have been accomplished on the airplane.

In addition to these records, if the airplane is to be operated under instrument flight rules (IFR*), each ten days and each ten hours of flight, the Omnirange navigation equipment (VOR*) must be checked for accuracy and the results recorded in a permanent record along with the name of the person performing the test. This can be done by the pilot. Also every twenty-four calendar months, the altimeter must be checked for accuracy by an approved repair station,* and the airplane's static system checked for integrity. These two checks must be recorded in a permanent record form.

2. Documents which must be in the airplane:

Maintenance records must be available for inspection by the FAA on request, but are not required to be kept in the airplane. Certain records and documents, however, must be in the airplane at all times that the airplane is operational. These are:

a. Registration Certificate:

This is issued to the owner of the aircraft and must show his name and current mailing address.

b. Airworthiness Certificate:

This certificate merely states that the airplane will be considered airworthy as long as it is maintained in accordance with the appropriate maintenance procedures.

c. Operations limitations* or flight manual*:

An operations manual is required for most modern aircraft. The part number for this manual will be called out in the Type Certificate Data Sheets. The manual includes the weight and balance information, the empty weight of the airplane, the empty weight center of gravity, and the location of all seats, fuel and oil tanks, and weighing points measured from the aircraft datum. Any limitations regarding baggage compartment loads are included to warn the pilot of any possibility of loading the airplane in such a way that the center of gravity will fall outside of the allowable center of gravity range.

An equipment list is included in the operations manual which lists the model and the part number of all equipment approved for installation, with its weight and center of gravity.

Engine operating limits such as RPM, manifold pressures, and temperatures, are included, along with a notation of the minimum octane fuel allowed.

Some of the older aircraft do not require a flight manual. Placards and markings on the instruments are sufficient to provide the pilot with the minimum required information. The Type Certificate Data Sheets tell whether or not a manual is required, and, if not, what placards and marking are needed.

d. Radio station license:

The Federal Communications Commission* (FCC) requires all radio transmitters to have a station license. This license must be displayed along with the registration certificate of the airplane; but, unlike the airworthiness and registration certificates, is not transferable with the airplane at the time of purchase. It must be issued by the FCC to the new owner of the aircraft.

QUESTIONS:

8. For how long must an airplane owner keep the record of an annual inspection?
9. How long must major alteration and repair forms (Form 337) be kept?
10. How often must altimeters be checked for accuracy if the aircraft is operated IFR?
11. What four documents or records must be kept in the airplane at all times it is operational?
12. How can you know whether or not a flight manual is required for a particular airplane?

SECTION II:

CONSIDERATIONS IN BUYING AN AIRPLANE

Buying a new airplane can be a rather straightforward transaction as far as maintenance is concerned. The dealer usually has a well equipped shop with trained A&P s to provide service for the new equipment, and warranty coverage usually protects the buyer from expensive maintenance problems in the early stages of ownership. Knowledgeable salespersons usually help the pilot match his requirements to the equipment, and, as a result, the owner starts off on the right foot with his new airplane.

Used airplane transactions may not be quite so simple, and to prevent disillusionment after the sale, these considerations should be taken:

A. SUITABILITY OF AIRPLANE:

Be sure the airplane meets your requirements: There is no such thing as a "good deal" or a "cheap airplane" if the equipment does not suit your needs. How important are such complexities as retractable landing gear, and a constant speed propeller? Is the ruggedness of the airplane compatible with the field you will be operating from? Are the electronics capable of any work you will be doing, yet not so complex that electronic maintenance and down time will be excessive? These questions must be objectively faced.

B. REPUTATION OF THE SELLER:

Be sure you are dealing with a reputable person. You will want the history and condition of the airplane honestly represented. A fair price can best be determined by a careful price study of as many comparable airplanes as possible.

C. AVAILABILITY OF PARTS AND SERVICE:

If the manufacturer is no longer in business, spare parts are likely to be difficult to find, and, if available, may be premium priced because of their scarcity. Insurance rates may be unreasonably high, and for some aircraft, may not even be obtainable. Be sure to check this out before becoming the proud owner of an uninsurable flying machine.

D. APPLICABLE A.D. NOTES:

Check with the FAA or with an I.A. regarding any Airworthiness Directives against this particular model of airplane, engine, propeller, and any installed equipment. An I.A. can also warn you of any outstanding manufacturer's service bulletins that may apply to your prospective purchase.

E. AIRCRAFT RECORDS:

After you know what to look for, paperworkwise, carefully examine the aircraft records.

1. A.D. Compliance Record:

Be sure that all of the A.D. notes have been complied with, and that the records show who did the work and the way it was accomplished. A simple record of the A.D. number and the letters CW for "complied with" are **not** adequate.

2. Equipment list:

Everything listed in the equipment list as being installed should be in the airplane. If radios or other equipment not listed are installed in the airplane, there should be a Form 337 for the installation.

3. History of the Airplane:

Carefully study the history of the airplane. Any long periods of inactivity should be explained. Engines which have been overhauled at an unusual time may have been damaged, and this possibility should be investigated. Airplanes which have been operated in coastal regions should be suspect for corrosion.

It is most generally true that the better an airplane has been maintained, the better will be the records. The converse is also quite often true. If the records are missing or inadequate, the airplane should be looked over all the more carefully.

4. Required documents:

Be sure all of the required documents are in the airplane: A current Airworthiness Certificate; a Registration Certificate showing the airplane to be registered to the person from whom you are buying; a flight manual or operations manual with a complete and up-to-date equipment list; a weight and balance record; and a radio station license listing every transmitter on board.

F. TITLE SEARCH:

After you are pretty sure that you have located the airplane you feel is the one for you, it is a wise investment to have a title search conducted. There are several companies in Oklahoma City, Oklahoma, who search the records at the FAA Aircraft Registry section in that city and will give you this important information: The name of the actual registered owner of the airplane, the one who must sign the bill of sale; any liens or encumbrances against the airplane which might exist; and the name of the previous owner. The fees for this service are well worth the peace of mind you have in knowing that there is likely to be no one later on trying to claim ownership of your airplane.

G. AIRCRAFT INSPECTION:

Without question, the wisest step in the purchase of a used aircraft is a careful inspection by a competent A&P technician. Airplanes are often sold with a "fresh annual." This sounds good, but can lull you into a false sense of security. For your own protection, you should know the A&P who is doing the work - one who will be able to thoroughly inspect the airplane and let you know exactly its mechanical condition. The details of a complete inspection are covered later in this book, but some of the more important things he will check for you are:

1. Signs of Corrosion:

Corrosion if it exists can be extremely costly. Hidden corrosion is usually the worst kind and requires a careful investigation by an experienced inspector. A few dollars "saved" by skimping here can make the purchase a bad investment.

2. Evidence of damage repair:

A properly repaired structure can be as good as the original, but sometimes repairs have not been properly made or recorded.

3. Hidden damage from rough landing or improper treatment.

4. Condition of the engine:

A differential compression check* and a careful examination of the oil filters and spark plugs tell an experienced A&P much about the internal condition of the engine. The external appearance, especially the baffles* and hoses, are indicative of the maintenance care the engine has had.

Considering the manufacturer's recommended time between overhaul, the hours on the engine, and its condition, the A&P should be able to give you a good idea of the amount of time you can expect before being faced with an overhaul. The fact that an engine has been recently overhauled is rather meaningless unless you know the integrity and skill of the person having done the work, exactly where it was done, and why.

5. Covering:

If the airplane is fabric covered, a fabric test is an absolute must. The A&P will make this test in that part of the fabric most likely to be weakened. The possibility of a re-cover job in the near future may have a pronounced effect on your decision to buy that airplane.

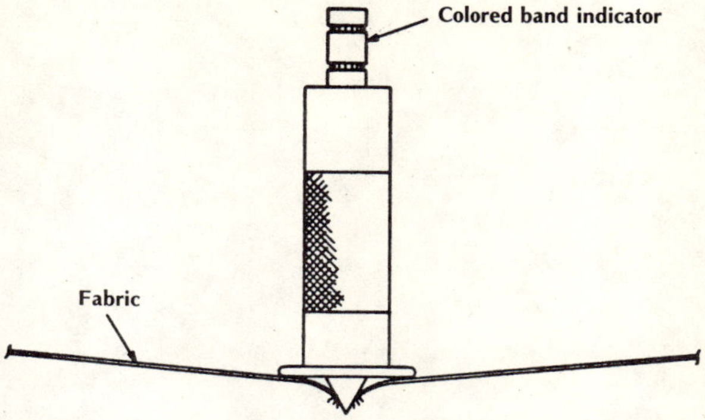

A fabric punch test is not a quantitive measure of the fabric strength, but does indicate the general condition of the covering.

- Figure 5 -

H. THE SALE:

After you are satisfied that this is the airplane for you, the sale is consummated. A sales contract is drawn up and executed between you and the seller, and an application for registration, FAA Form 8050-1, is filled out. The white and green copies are sent to the FAA Aircraft Registration Branch in Oklahoma City, Oklahoma, along with a copy of the bill of sale and a release of any liens that you may have paid off. The pink copy of the application is placed in the airplane where it must be kept until the registration certificate is received.

If the airplane is being financed, the lender will have made the title search, and along with the bill of sale he will record his lien against the airplane.

In addition, you, as the purchaser, must make your own application at this time with the Federal Communications Commission for an aircraft radio station license.

QUESTIONS:

13. How can you be sure that the airplane you are buying has a clear title?
14. Is a "fresh annual" inspection given an airplane before it is sold an assurance that it is in top condition?
15. What must be done with the application for registration of an aircraft at the time of sale?
16. Is the radio station license transferable to the new owner of an airplane?

SECTION III: AN INTRODUCTION TO AIRCRAFT SYSTEMS

Warren truss fuselage made of chrome molybdenum steel tubing, covered with cotton, linen, or Dacron fabric.

- Figure 6 -

A. AIRCRAFT STRUCTURES:

The requirements for an efficient aircraft structure have taxed the ingenuity of aircraft engineers since the days of the Wright brothers and Glenn Curtiss. The structure must be sufficiently strong to withstand the flight loads and landing stresses, and it must be light weight so that more pay load can be carried. It must be streamlined to provide more efficient air passage and allow higher airspeeds; it must be easy to manufacture and simple to repair. Its materials must withstand radical temperature changes without excessive dimensional change. It must absorb vibration without cracking and must tolerate all kinds of environmental conditions without corroding or deteriorating.

The changes in our technology in the past several decades have forced aircraft structure to change from that which required much highly skilled craftsmanship to the more easily mass produced sheet metal structure.

1. Types of structure:

Airplane structure falls into two basic categories: truss* and stressed-skin*. Although the vast majority of modern aircraft are of the sheet metal, stressed-skin type, truss fuselages* and wings were the first to be used and are still extant in a few aircraft.

a. Truss:

Rigid members designed to carry tensile or compressive loads form the framework of some aircraft. Around this basic truss, formers* are attached, to give the component its aerodynamic* shape; streamlined, in the case of the fuselage, and shaped to produce lift, for the wing.

Most truss fuselages use a Warren truss, figure 6, in which the cross members carry both tensile* and compressive* loads. Fuselage trusses are normally welded up of thin-wall chrome molybdenum steel tubing. Attached to the tubing to produce the desired streamlined shape are wooden or light metal formers, and thin metal or wood strips called stringers*, running the length of the fuselage to fill in the shape of the formers. The entire structure is covered with a strong fabric of cotton, linen, or one of the synthetic fibers such as Dacron*.

Wing structures are usually of the Pratt truss, figure 7. In this truss, the lengthwise members called spars* are separated by metal tubes called compression struts* which carry only compressive loads.

A pratt truss uses two spars to carry the main loads of the wing. They are separated by compression members held in place by drag and anti-drag wires.

- Figure 7 -

Running between the spars from one compression strut to the next, in the form of an "X" across the truss, are hard steel wires called drag* and anti-drag* wires. The combination of spars, struts and wires is held tightly together by the tension on the wires. Wing ribs, to give the wing its shape, are slid over the truss and attached to the spars. A thin sheet aluminum wrap-around leading edge* and a somewhat heavier gage trailing edge* complete the structure. The entire wing is covered with fabric such as that used on the fuselage.

The truss structure has the advantage of relatively high strength for its weight, and ease of fabrication and repair have, in the past, been factors in its favor.

b. Stressed skin:

The chief disadvantage of truss construction lay in the requirement for two separate structures: one, the truss, to provide the structural strength; and the other, the formers and stringers, called superstructure,* to give the desired aerodynamic shape. The availability of strong, thin sheet metal has made it possible to build a stressed skin structure without a truss, which has the required strength along with the smooth aerodynamic shape needed for an airplane. In the development of stressed skin structure, several types have been used, but today there are basically two which find use in small private aircraft.

[1] Monocoque*:

The monocoque, or eggshell type construction has, for a long time, been known as an ideal type for aircraft; but only since the advent of strong, thin aluminum sheets and production methods for forming these sheets, did this type of construction really become popular. A typical light aircraft fuselage consists of former rings to which thin sheets of aluminum are riveted. It is the shape of the skin that gives it its strength. To demonstrate the principle of a monocoque structure: If both lids were cut from a tin can, the can stood on its end, and a board placed across the other end, the thin steel wall of the can could bear the weight of a full grown man. All of the load is supported by the way the can is designed. If you were to take the can in your hand and squeeze it on its side, you could easily crush it. This is because it is not designed to withstand stresses in this direction. Monocoque structures cannot tolerate any dents which destroy the shape, as it is the shape which actually allows the metal to carry all of the loads.

The aluminum alloy used in most aircraft structure is

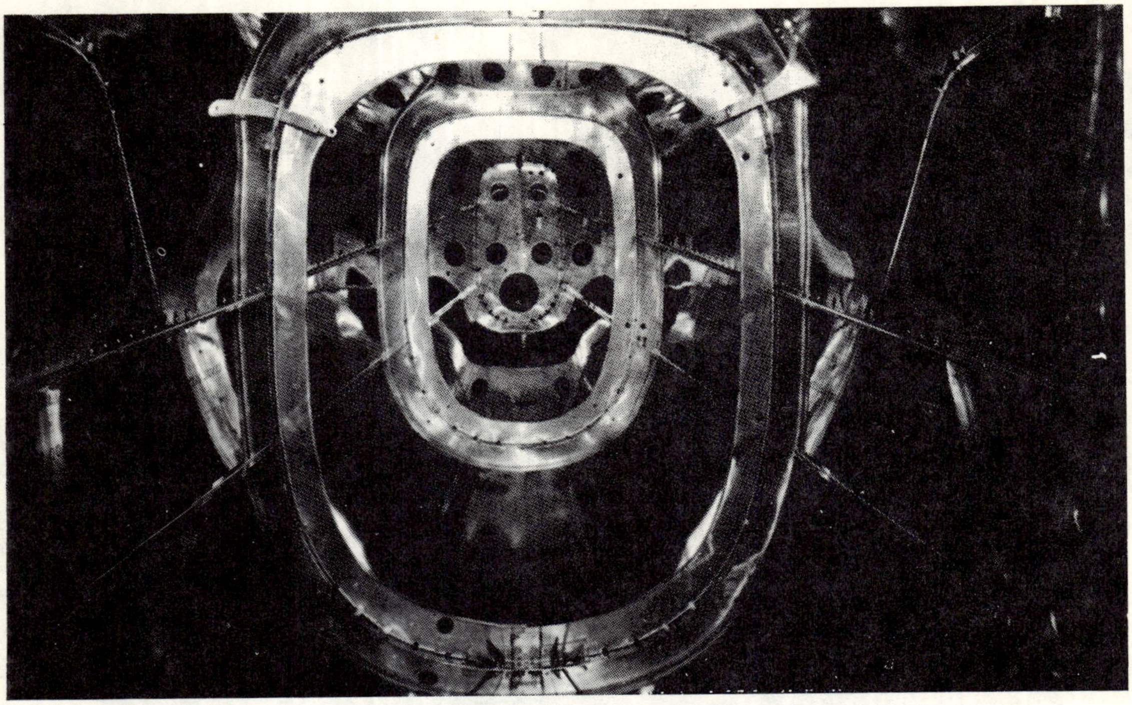

A monocoque structure carries all of the stresses in the outer skin. The only framework is the rings which give the structure its shape.

- Figure 8 -

amply strong, more than 60,000 pounds per square inch in tensile strength. The problem in its use lies in its rigidity or resistance to bending. For this reason former rings, stiffeners* and doublers* are used in strategic locations to help the skin keep its shape and accept the loads in ways which favor its strength. One common way to stiffen an area is to use flanges bent up at the edges of sheets, figure 9, or channels bent up of thin sheet metal riveted to the skin, figure 10a.

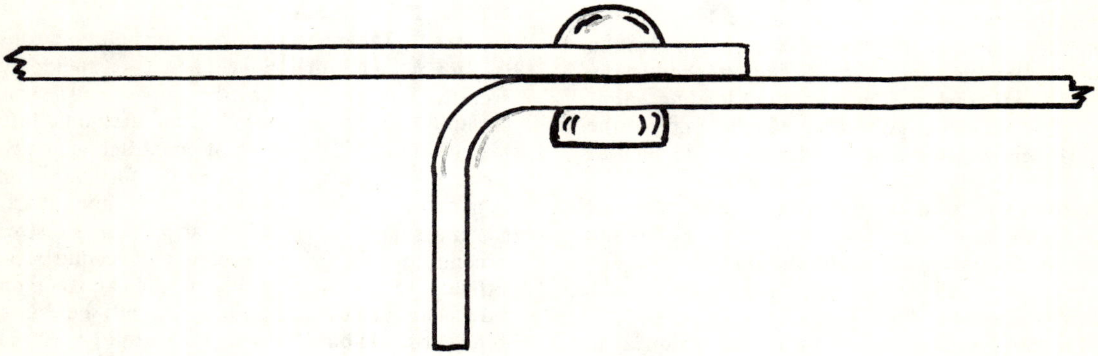

Simple monocoque structure is stiffened by bending a flange on the edge of the sheet.

- Figure 9 -

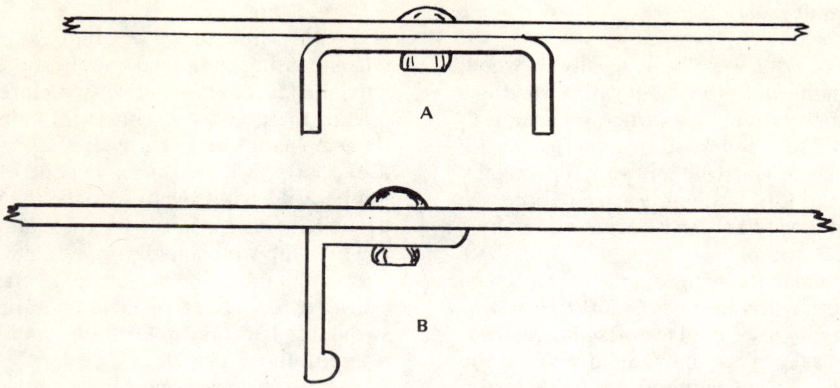

Channel [A] or bulb angle [B] is used to stiffen the skin in a monocoque or semi-monocoque structure.

- Figure 10 -

[2] Semi-monocoque:

Pure monocoque structure must be limited to the smaller, lightweight aircraft. For structures requiring more rigidity than can be provided by a monocoque, a modification known as semi-monocoque is used. In this, the former rings are connected by a substructure made up of stringers, usually bulb angle extrusions, figure 10b. The skin is riveted to the former rings and stringers, which results in a structure far more rigid than without such a framework.

Most stressed skin wings are semi-monocoque. The ribs form the shape, and the stringers add the necessary stiffness. Rather than simply a front and a rear spar, as in truss wings, some semi-monocoque wings use several smaller spars.

QUESTIONS:

17. What is the major disadvantage of a truss type aircraft structure?
18. Why is it of extreme importance that a monocoque structure not be dented?

---

2. Structural Materials:

a. Metallic materials:

[1] Aluminum alloys*:

The most commonly used structural materials in modern private aircraft are the aluminum alloys. These metals have the advantage of high strength for light weight, they resist attack from the environment fairly well when properly protected, and are available in such quantity that their cost is reasonable. Aluminum alloys are easy to form and may be heat treated to very high strength.

Pure aluminum is soft and not very strong, so it has little use as a structural material; but when aluminum is alloyed with small portions of copper, manganese, and magnesium, its strength goes up from about 13,000 pounds per square inch to around 68,000 psi tensile strength. Copper alloyed aluminum is known as 2024 and in its heat treated state has the suffix -T3 added.

When aluminum is alloyed, it gains in strength but loses in its ability to resist corrosion.* Pure aluminum forms a dense oxide film on its surface which resists further chemical action with the air or with moisture. Aluminum alloys, on the other hand, are quite susceptible to corrosive action when covered with airborne moisture. So that we can have our cake and eat it too, a piece of the structural alloy is coated on both sides with a thin layer of pure aluminum. This is known as clad aluminum alloy. In the rolling mill, slabs of pure aluminum are placed on the ingot of aluminum alloy and rolled together until they are the desired thickness. In this way, the pure surface metal has actually become a part of the base alloy sheet. There is naturally a small sacrifice in strength for this corrosion protection, but it is considered a worthwhile trade.

[2] Magnesium alloys:

It is often desired to use a metal lighter than aluminum for making a part. Magnesium alloys are used in this case. Movable control surfaces and wheels are the two components most often made of this light weight metal. As for the inevitable compromise: we pay for the weight saving by corrosion susceptibility and a tendency toward cracking.

The pilot should be especially watchful of magnesium structure for cracks and the onset of any corrosion. These must be immediately dealt with.

[3] Steel:

Truss type fuselages are welded up of thin wall, chrome molybdenum (chrome moly) tubing, a very tough alloy of steel. Aircraft hardware such as nuts and bolts are generally made of nickel steel. Certain highly stressed parts such as spring steel landing gears, engine crankshafts and valves, are made of special alloys specifically suited for these applications.

Steel has a tendency to rust* or corrode when exposed to oxygen in the atmosphere, and since many steel parts are used as fasteners and cannot be painted, they are protected from rust by plating their surface with a thin coating of cadmium. Cadmium is more susceptible to corrosion than steel and will be the metal eaten away when corrosion occurs - thus protecting the steel.

QUESTIONS:

19. What is one disadvantage of aluminum alloy as a structural material?
20. What is meant by clad aluminum alloy?
21. What are two disadvantages of magnesium alloy for aircraft structures?
22. What is the purpose of cadmium plating on aircraft hardware?

---

b. Nonmetallic materials:

[1] Wood:

Early day aircraft were built mostly of wood, and because of its high strength to weight ratio, wood has fought a good battle for survival as a structural material. Now with labor costs so high, and wood structure requiring so much more labor in its construction and repair than metal, wood is becoming less frequently used. There are basically three types of wood likely to be encountered by a pilot: solid, laminated,* and plywood.*

Solid wood may be used for wing spars of truss type wings. The main wood for this purpose is Sitka spruce.

Laminated wood may be used in place of solid wood for spars, primarily for the sake of economy. Solid spars of the length and degree of perfection required are quite costly, so a spar built up of shorter, thinner pieces glued together with the grain all running the same direction have proven more economical and just as strong. Wooden propellers are carved from a blank made up of birch planks laminated together.

Plywood is made of thin sheets of wood called veneer, glued together with the grain of alternate sheets placed at 45 degrees or 90 degrees to each other. This makes a very strong wood, and one which can be formed and shaped much as metal. Birch plywood is used for reinforcements, and mahogany plywood for wooden stressed skin. Plywood skins are normally covered with lightweight fabric and produce exceptionally smooth, strong, and durable surfaces.

[2] Fabric:

Truss type aircraft structure must be covered with a cloth fabric, shrunk tight and treated to make it waterproof. Long staple, Grade A cotton is the standard fabric. It is just about as thick as a good quality dress shirt, and is made up of 80 threads per inch in both directions. It is extremely strong, having a pull test of 80 pounds per inch. British linen is an acceptable substitute for cotton, provided it meets the requirements the FAA has set up for its manufacture. It is a little more difficult to work with but has longer life than cotton.

In recent years a new covering material has been developed which has had excellent reception among airplane owners and mechanics alike. It is Dacron fabric, sold also under such proprietary names as Ceconite and Eonnex. Airplanes which left the factory covered with cotton may be re-covered with this material by using the authorization provided by a Supplemental Type Certificate (S.T.C.), issued to the seller of the fabric. Dacron, a non-organic material, has a much longer life than either cotton or linen, and is considerably stronger than either. The Dacron used is a special fabric which has not been shrunk in the textile mills. It is put on the structure quite loose, then shrunk with an electric iron or a heat lamp.

Fiberglass cloth is an approved covering material for aircraft when applied according to an S.T.C., but it has several characteristics, including a very intolerant application procedure, which makes it less popular than Dacron or the older organic fabrics.

At first thought, a covering material with the longest service life would seem to be the best investment. This is not necessarily so, however, because the structure beneath the fabric was not designed to go without inspection for years on end. It should therefore be uncovered and thoroughly inspected for rust or other corrosion within a reasonable time period.

[3] Plastics:*

The term "plastics" is such an all encompassing one that its applications run the gamut from windshields to the entire structure. Basically, aircraft plastic materials are considered under two categories: thermoplastic* and thermosetting.* Thermoplastic resins are those which soften when exposed to elevated temperatures and reharden once the heat is removed. The best example of a thermoplastic material in common use is the acrylic* windshield and side windows found in most light airplanes. This material is transparent, somewhat brittle, especially at lower temperatures, but lightweight, strong, and so easy to form that it enjoys almost universal usage in these places.

Thermosetting resins comprise the vast majority of the plastic structural materials used today. A thermosetting resin is one that once set, usually by chemical means, will not soften when heat is applied. This resin is usually reinforced with some type of fabric or fiber mat material. Control pulleys of this material are often reinforced with linen fabric; electrical insulators, with paper; and structural forms such as wing tips, wheel pants, propeller spinners, radomes,* and even some landing gear are reinforced with glass fibers, either woven into a cloth, or formed into a random lay mat.

Fiberglass reinforced plastic resin is used to make such parts as this wheel pant.

- Figure 11 -

Polyester, epoxy, or phenolic resins are used to bond the fibers together and form a rigid, strong component.

[4] Finishes:

Paint and finishing materials are such a huge field in themselves that we must consider the various applications to see what is available:

[a] For wood structure:

Most wood structure not visible from the outside of the airplane is covered with a good coat of spar varnish.* This hardens by chemical action within the varnish, and produces a finish highly resistant to water.

[b] For fabric surfaces:

Fabric which covers plywood and that which provides the skin of the airplane itself is treated with

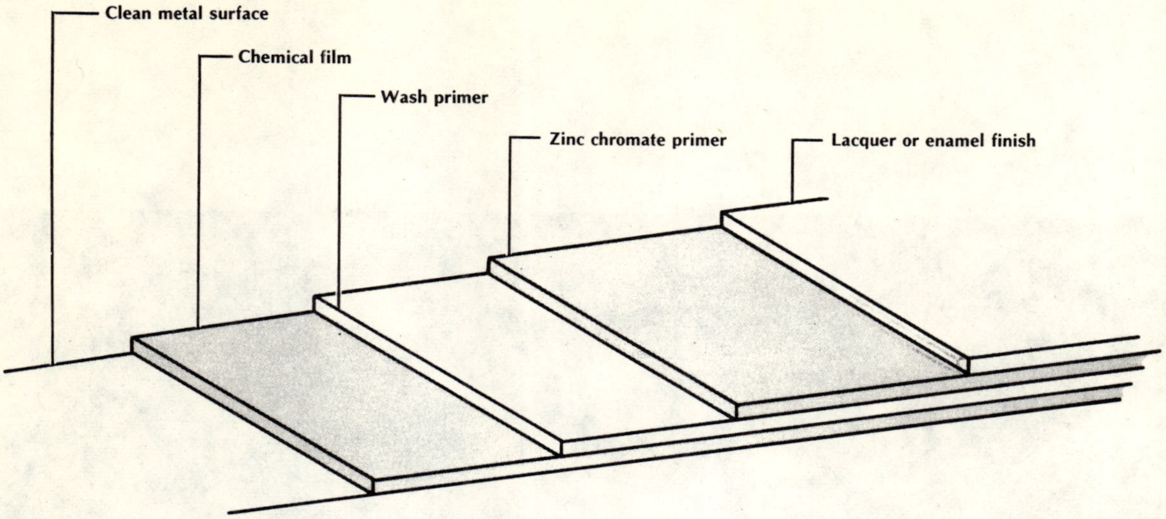

For maximum durability, the surface of a metal aircraft is treated with a conversion coating, and then a wash primer. To this prepared skin, an epoxy primer and a polyurethane enamel finish may be applied. Most high-volume production finishes have only the wash primer and an acrylic lacquer.

- Figure 12 -

material known as dope.* Nitrate dope is made of cotton fibers dissolved in nitric acid and mixed with solvents and plasticizers* to reduce the brittleness of the film. The big disadvantage of nitrate dope is its almost explosive burning rate. To reduce the fire hazard, butyrate dope is now almost universally used. It consists of fibers dissolved in acetic acid and mixed with butyl alcohol as a solvent. Plasticizers are added to make the film more resilient.

Both nitrate and butyrate dope produce a clear, transparent film on the fabric. To prevent the sun damaging the fabric and the bonding coats of dope, aluminum powder is added to clear dope for those coats applied just before the color. Aluminum powder consists actually of tiny flakes of aluminum metal. These almost microscopic flakes are mixed into the clear dope and sprayed onto the surface. Before the dope dries, the surface tension of the liquid causes these flakes to leaf out and form a solid covering of this extremely thin aluminum over the surface. This coating is opaque and prevents the ultraviolet rays of the sun damaging the fabric or the dope.

[c] For metal structure:

/1/ Primers:

The adherence of the finish coat of paint to any metal surface depends on the preparation for that finish. Steel tube structure must be protected by a finish that will not only protect the metal from rust or corrosion but will not be softened or lifted by the dope used on the fabric. Zinc chromate is the most universally used aircraft primer. It is an inhibiting primer, in that the resins which hold the zinc chromate allow a very slight penetration of water into the finish. This water causes chromate ions to be released, inhibiting the formation of any corrosion. Besides limiting corrosion, zinc chromate primer provides a good bond between the metal and the finishing lacquer or enamel.

An epoxy primer may be used on steel tube structures subject to exceptionally corrosive environments, such as agricultural aircraft or seaplanes. It is a two-part chemical in which a catalyst* causes the resin to set up and form an impervious film over the metal.

A wash primer is used in the preparation of aluminum surfaces for some of the new acrylic finishes. This, like the epoxy, is a two part primer, but instead of simply forming a film over the surface, the wash primer etches the metal, eating the surface away to provide microscopic roughness for better adhesion of the top coats of finish.

/2/ Surface finishes:

Metal surfaces may be finished with either an enamel or a lacquer finish. Lacquer is made of pigments held in a nitrocellulose film base, along with plasticizers to prevent cracking. They do not fill as well as do enamels. Acrylic lacquer is one of the more popular lacquers because of the durable film it leaves.

Enamels have their pigment suspended in a varnish base. They tend to form a smoother finish than lacquer, and they hide surface imperfections better. A new and popular finish for metal airplanes is polyurethane enamel. This is a catalyzed material which forms a hard, tight, and durable finish which retains a high gloss without waxing or polishing.

QUESTIONS:

23. What is the difference between a laminated wood and plywood?
24. What is the difference between a thermoplastic and a thermosetting material?
25. Why is aluminum pigmented dope used beneath the colored coats of dope?
26. How does zinc chromate primer protect a surface from corrosion?
27. What is the purpose of a wash primer for a metal aircraft finish?
28. What is the difference between enamel and lacquer?

The bungee shock cords stretch to smooth out landing and taxi roughness.

- Figure 13 -

B. LANDING GEAR:

There is no single part of an airplane subject to as much hard usage as the landing gear, and the requirements for its design and construction are some of the most stringent in the entire airplane. It must be capable of withstanding all of the impact loads a hard landing subjects it to and must be able to take side loads imposed during a cross wind landing. It must support a heavily loaded aircraft for its take-off run, sometimes over rough fields. It must have a minimal frontal area, if it is a fixed gear; or if retractible, it must function every cycle with the least possible chance of failure. Mud, water, ice, rocks and sand are continual enemies of the landing gear. To better understand this important system, we will study its components:

1. Shock struts:

Devices to handle the landing impact and taxi loads may be considered in one of two categories:

a. Non-absorbing types:

You may occasionally encounter bungee shock cords,* such as were used on many of the earlier Piper aircraft up through the Tri-Pacer series. They are composed of hundreds of rubber filaments about 1/16'' square, formed into a ring about eight inches to a foot across, and in bundles up to about 7/8'' diameter. These rings are encased in a loose weave cotton braid which allows for stretching. The ring is stretched around the landing gear so that when a load is imposed on it, it stretches out. On later model aircraft using this type landing gear, a hydraulic snubber* prevented the rubber snapping the gear back too rapidly.

Cessna single engine aircraft have, for years, used a spring steel landing gear that actually has little shock dissipating action, but receives the landing shock and gives it back - usually in such a manner as not to cause the airplane to rebound back into the air. Two forms of this gear are in use: the flat gear, which flexes purely in bending; and the later model, round in the cross section, which flexes in torsion* as well as in bending. Other aircraft use this form of gear and find it very satisfactory from both the operational and maintenance points of view.

The spring steel landing gear is not an actual shock absorber. It stores the landing energy and returns it as the airplane rolls out after landing.

- Figure 14 -

b. Shock absorbing types:

All large aircraft and almost all smaller ones not using the simple spring type gear use a form of air-hydraulic shock absorber, commonly called an "oleo"* strut. On fixed gear airplanes, this attaches directly to the wing structure, or, in some cases, to the fuselage. Nose gear cylinders are often built into the engine mount structure. The piston holds the axle for the wheel and the brake mechanism, while a set of torsion links,* or scissors, holds the piston in the cylinder, allowing it to slide in and out, but preventing its rotation. It also prevents the piston coming out of the cylinder. The piston is sealed in the cylinder with either O-rings* or chevron seals. When the weight of the airplane is on the gear and the strut is collapsed, the piston is all the way in the cylinder.

An oleo strut absorbs shock as the piston, attached to the wheel, moves up into the cylinder which is attached to the airframe.

- Figure 15 -

The cylinder is filled with aircraft hydraulic fluid when the strut is collapsed, and then it is inflated with high-pressure compressed air or nitrogen until the piston is exposed the amount specified in the maintenance manual, usually about three inches.

The cylinder is divided into two chambers, and when the strut is extended as while flying, all of the fluid is in the lower chamber. On landing, the piston is pushed back into the cylinder, and the oil transfers into the upper chamber. It is restricted by the small hole, and much of the landing energy is changed into heat in the fluid. Taxi loads are taken up by the compressed air after the initial impact is absorbed by the fluid.

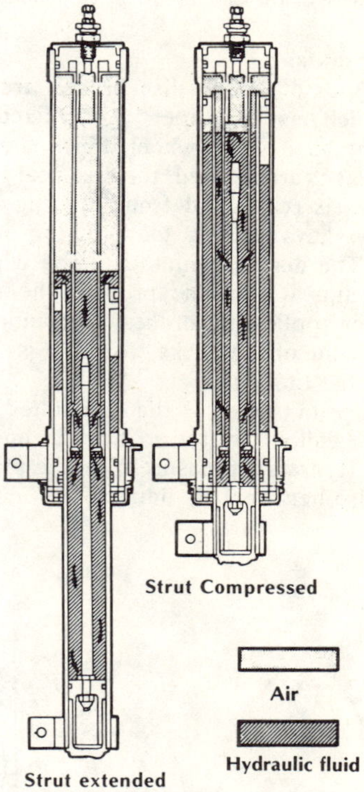

Internal construction of an oleo shock strut.

- Figure 16 -

2. Wheels:

Some of the earlier airplane wheels were single piece units similar to those on the automobile; others had one removable rim, as do some large truck wheels. Now, the most widely used wheel is a two piece split wheel, whose halves are held together with through bolts.

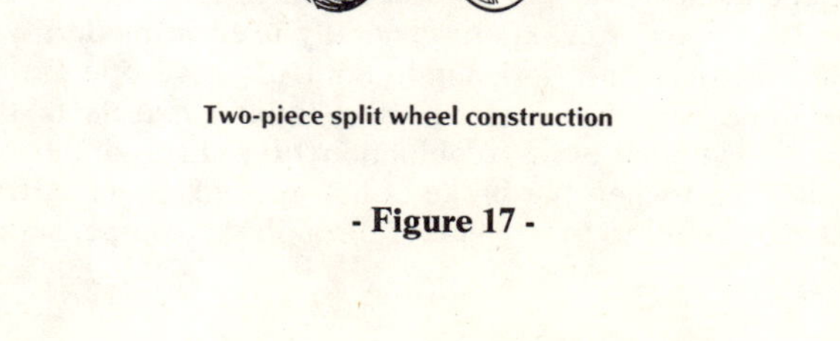

Two-piece split wheel construction

- Figure 17 -

This type of wheel can use either a tube type or a tubeless tire. If a tubeless tire is used, the two wheel halves are usually sealed with an O-ring to prevent the loss of air.

Wheels may be made of aluminum or magnesium alloy.

3. Brakes:

Aircraft brakes are almost all of the hydraulic actuated, nonenergizing type. "Nonenergizing" means that the weight of the airplane does not help apply the braking action. These brakes are usually of the disc type, using either a single disc, or, in the case of larger aircraft, multiple discs. Since pilot performed maintenance will most generally be done on the single disc brakes, these are the ones which will be discussed in this book.

a. Wheel units:

Two types of single disc brakes are in common use, one of which has the caliper* rigidly fixed in a housing attached to the axle of the wheel. Since something must move when brakes are applied, the disc floats in and out in the wheel, but is restrained from rotating by hardened steel disc drive keys, or by a toothed ring into which the disc splines. The disc is centered in the wheel by small spring steel clips which also prevent the disc rattling. When the pilot applies the brakes, hydraulic fluid moves the piston and one of the pucks,* or linings, clamping the disc between the linings.

The other type of brake has the disc bolted rigidly to the wheel and the caliper is free to float in and out on two anchor bolts. Hydraulic pressure from the master cylinder clamps the disc between the linings.

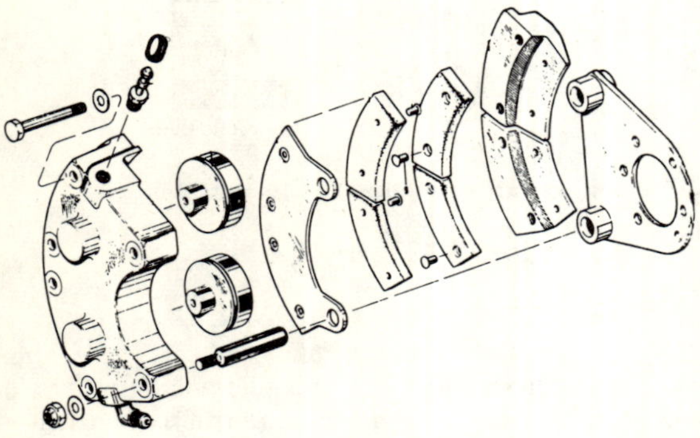

Caliper of Cleveland brake. The disc is bolted to the wheel and the caliper moves in and out on the two anchor bolts.

- Figure 18 -

b. Master cylinders:

Large aircraft having complex hydraulic systems may use system pressure to apply the brakes, but most small aircraft use hydraulic pressure generated in independent foot-operated brake master cylinders. There are two types of cylinders generally used in modern aircraft: vertical and horizontal. Both of these operate on the same principle. The wheel cylinder connects to the master cylinder by a combination of rigid and flexible tubing, and when the brake is not applied, there is free passage between the cylinders through the compensating port*. Any expansion of fluid due to heat will be taken up by the reservoir, eliminating brake drag. Also, if there is any loss of fluid, it can be replaced from the reservoir each time the brakes are released. When the pilot first applies the brakes, the compensating port closes; the piston in the master cylinder then pushes fluid into the wheel cylinder, causing the caliper to clamp the disc.

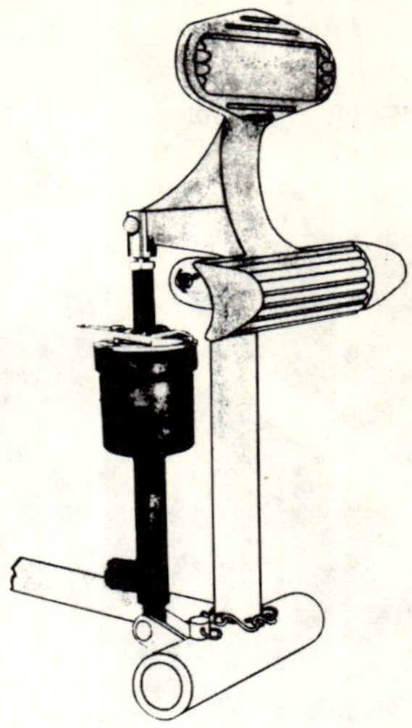

Typical vertical brake master cylinder

- Figure 19 -

Parking brakes are applied by mechanically locking the piston in the applied position. Smaller aircraft do not usually have special provisions for thermal expansion of the fluid when the brakes are applied, as this is normally such a small amount.

Some aircraft use only one master cylinder for both brakes. This type of installation uses a "T" in the line from the master cylinder, and both wheel cylinders are supplied with fluid at the same time. Independent braking cannot be done, of course, with this type of system.

Brake reservoirs are located either on the master cylinder itself, or they may be located remotely, usually on the firewall where they can be checked on a preflight inspection. Almost all of the general aviation piston engine aircraft use MIL-H-5606 mineral base hydraulic fluid in the brakes.

QUESTIONS:

29. What cushions the shocks produced in taxiing when an oleo shock strut is used?
30. What absorbs the landing impact with an oleo shock strut?
31. What kind of fluid is used in an oleo shock strut?
32. How are wheel halves sealed to prevent air leakage when a tubeless tire is used?
33. What is the purpose of the compensating port in a brake master cylinder?

---

4. Tires and tubes:

There are eight types of tires designed for aircraft, each type having its own application and requirements. The aircraft discussed in this book almost all use type III, smooth contour tires, which may be had in several ply ratings and tread designs. The most common tread is simply a series of circumferential grooves which has proved quite effective for braking with varying runway conditions.

Goodrich

The most commonly used tire tread design for light aircraft is the ribbed tread.

- Figure 20 -

One or two bundles of steel wire embedded in rubber provide an anchor for the ply fabric and a firm mounting surface for the wheel. These are called beads. Chafers or chafing strips add protection over the ply fabric, preventing damage in mounting or de-mounting and minimizing deterioration from overheating by brake action. A heavy layer of specially compounded rubber has the tread design moulded into it.

The tread gets the most wear, as the wheel must accelerate from zero to the touchdown speed in a very short distance, with this accelerating force applied to the tread. It is usually worn off before the carcass has much wear, and for this reason aircraft tires are quite often retreaded. Retreading is done with the sanction of the FAA as long as it is done by a repair station approved for this work. Any retreaded tire must have the name of the station doing the work permanently marked on the tire, as well as the date the work was done, and the number of times the carcass has been retreaded.

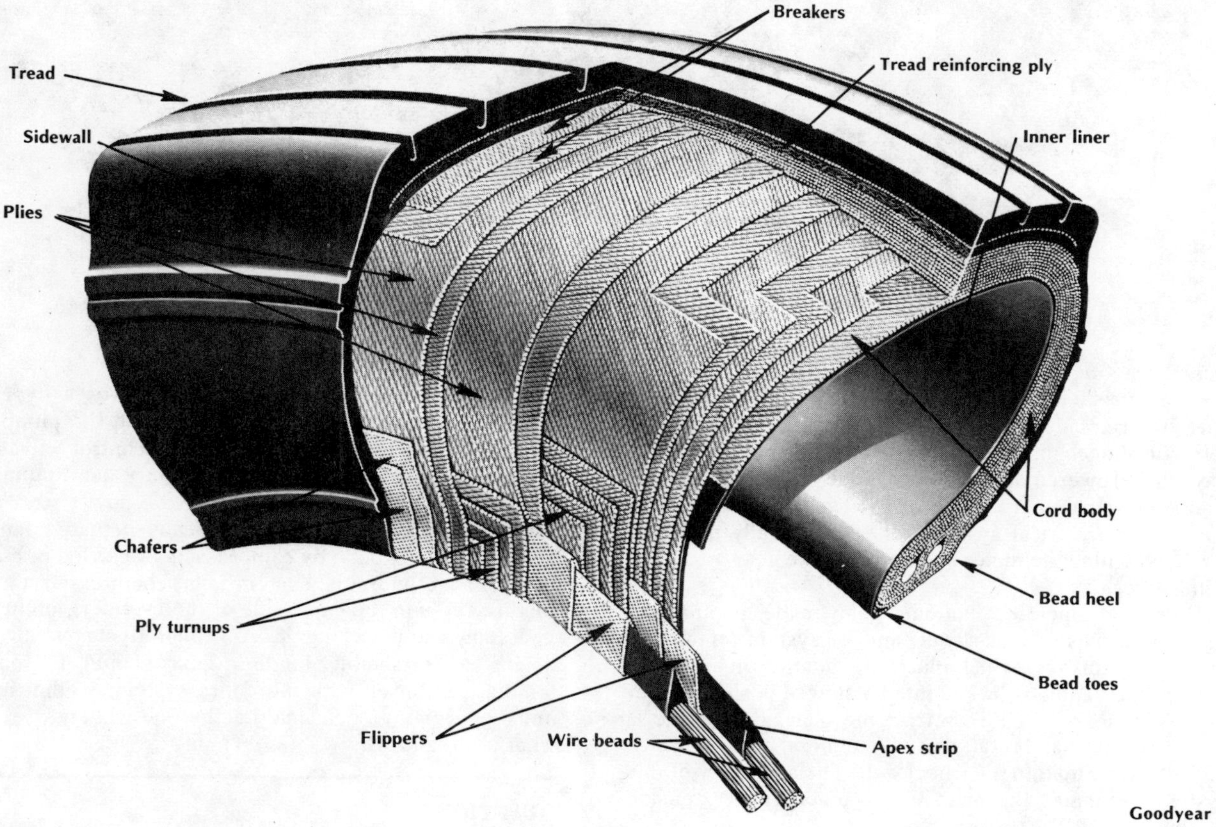

Typical tire construction

- Figure 21 -

As cruising speeds have increased, landing gears have been retracted into the airframe to lower the parasite drag.

- Figure 22 -

5. Retracting System:

As airplanes reach higher speeds, the effect of parasite drag* increases, and the additional weight of the retracting mechanism is not an unreasonable penalty to pay for the lowered air resistance. Electric motors, driving chains or worm gears, raise and lower the wheels. As a backup for electrical systems, there is usually provision for de-clutching the motor and lowering the gear by a hand crank lever.

The hand operated landing gear made popular by Mooney consists of a lever and a system of purely mechanical linkages which places the gear down and locks it when the handle is locked into its stowed position under the instrument panel. To retract the gear, the handle is simply released and rotated to its horizontal position which pulls the gear up into the wheel well. The simplicity of this system eliminates the need for an emergency extension system.

Hydraulic landing gear systems have been used extensively by the larger transport aircraft, and they are quite popular on smaller aircraft also. Most small aircraft do not have a complex hydraulic system, so provisions are made for a simple system which will adequately do the job. One approach to the simplified system is the hydraulic power-pack. Here, an electric motor drives a hydraulic pump. Incorporated in the same unit with the pump and motor are the reservoir and whatever control valves are necessary. Systems in which an auxiliary hand pump are required usually have this pump in the power pack also. Emergency extension of hydraulically actuated landing gear is normally done by compressed air carried in a steel bottle. If the hydraulic system fails, compressed air may be directed into the down side of the cylinder, forcing the gear down and locking it. Any pilot flying an airplane having a retractable landing gear should thoroughly familiarize himself with the entire system, so that in the unlikely event of a gear malfunction, he will know exactly what to do and **why**!

QUESTIONS:

34. Are retreaded tires approved for use on aircraft?
35. What type of emergency system is normally used with:
    a. Electrically actuated landing gear?
    b. Hydraulically actuated gear?

C. CONTROL SYSTEM:

1. Primary Aerodynamic Controls:

a. Ailerons:*

Ailerons for light aircraft are normally actuated from the control wheel cables. A sprocket is mounted on each control shaft and a roller chain links the two wheels, or a pulley and cable arrangement may connect the wheels directly to the aileron bell cranks. A balance, or carry through, cable joins the ailerons to make one complete loop of the control cable.

The differential action of the ailerons which minimizes the effect of aileron yaw is achieved by the use of a bell crank,* such as figure 23. Because of the geometry of this arrangement, the aileron moving up is deflected more than the one moving down.

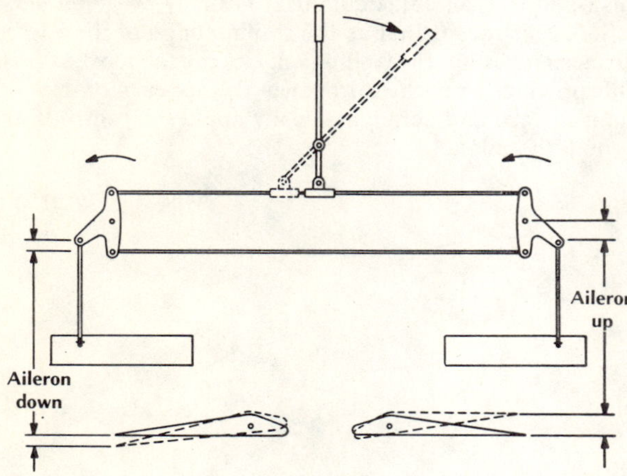

Differential aileron movement is obtained by the use of offset bellcranks.

- Figure 23 -

b. Elevators:*

Elevators are the movable horizontal surface of the empennage* and are actuated by in and out movement of the control wheel to rotate the airplane about its lateral axis. Elevators may be linked to the wheel by cable, or by a long hollow aluminum tube known as a push-pull rod. Either system attaches to the surface through a bell crank in the tail cone of the fuselage.

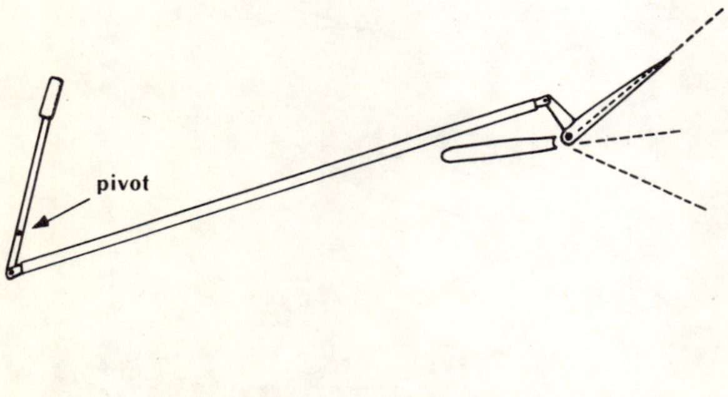

A rigid aluminum tube moves the control both up and down and is often used instead of control cables.

- Figure 24 -

c. Stabilator:*

Rather than the fixed horizontal stabilizer and movable elevator, some modern light aircraft make use of a completely movable horizontal tail surface. This stabilator serves the function of both of the horizontal surfaces and may be controlled in the same way. There are two differences between this type of control and the more conventional two piece surface. The stabilator pivots about a point somewhere near one third of its chord. To prevent the control being overly sensitive due to the large area ahead of the pivot, an anti-servo tab* is installed on its trailing edge.

As the trailing edge of the stabilator moves up, the anti-servo tab also moves up.

- Figure 25 -

The second problem inherent with a stabilator is flutter. This is prevented by use of a rather large counterweight on an arm which protrudes into the fuselage. This weight moves the center of gravity of the surface forward of the hinge point to prevent flutter.

d. "V" tail surfaces:

The "V" tail Beech Bonanza uses two surfaces rather than three in the empennage to provide both directional and longitudinal control. The functions of both elevator and rudder are achieved by a set of bell cranks, which allow the in and out motion of the control wheel to operate the movable surfaces together and the rudder pedals to operate them differentially.

The V tail of the Bonanza serves the same purpose as the more conventional three-piece tail.

Beech

- Figure 26 -

e. Rudder*:

The conventional rudder is operated directly by cables from the rudder pedals to the surface. There is no complete loop of cable, as there is with the ailerons. The rudder pedals usually work the nose gear steering mechanism also, through a push-pull linkage to a steering bell crank, or whiffletree.*

A common method of accomplishing nose gear steering for retractable gear aircraft is for the steering bell crank to push against rollers on the nose gear strut. When the gear is retracted, the rollers rotate away from the bell crank and in no way disturb the rudder action.

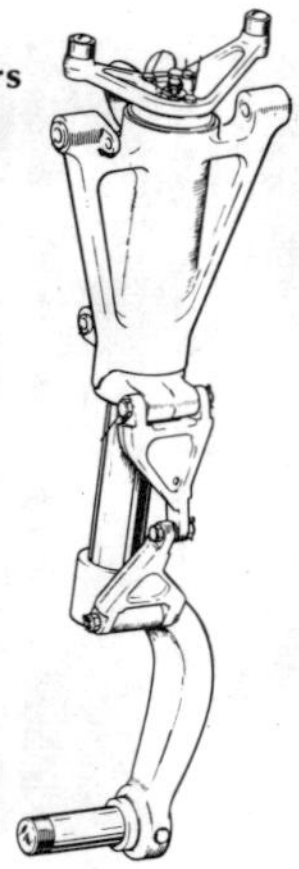

Nose gear steering on a retractable nose gear is accomplished by the steering mechanism pushing on rollers at the top of the strut.

- Figure 27 -

2. Secondary aerodynamic controls:

a. Wing flaps*:

Wing flaps for most light aircraft are of either the simple or of the Fowler type.

Simple flaps are hinged about a point and lowered by rotating them about this point. They are normally operated by hand with a cable from the flap control handle. Intermediate positions between full up and full down may be obtained by locking the handle into notches in its mount. A button on top of the handle releases the lock.

Fowler flaps, instead of pivoting about a point, slide out from the wing on a track. These flaps are pulled up and down by cables, actuated either by a hand lever or by an electric motor-driven jack screw.

Fowler flaps move out of the wing on tracks.

- Figure 28 -

b. Tabs:

Movement of the primary control rotates the airplane about one of its three axes. For a steady hands-off cruise, a tab attached to the surface deflects to change the aerodynamic shape of the control and allows the airplane to balance without the pilot having to hold pressure. There are three basic types of movable tabs used on modern control surfaces:

[1] Trim tab:

A simple tab, adjustable from the cockpit, may be moved a fixed amount to change the balance condition of the airplane. This is the most simple and is used where the least amount of aerodynamic assistance is required.

[2] Servo Tab:

By simply moving the point of attachment for the push-pull rod, an adjustable trim tab may be given servo action. This means that as the trailing edge of the control surface moves up, the trailing edge of the tab moves down. This provides a pressure drop over the upper portion of the main surface and aerodynamically applies a control force to help the pilot.

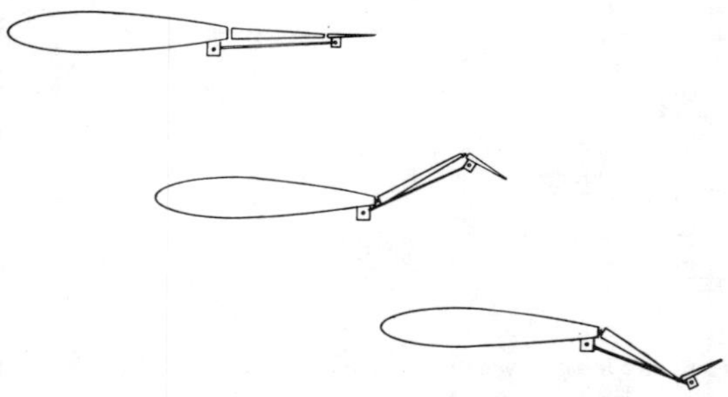

Servo tabs move in the direction opposite the primary surface and aid the pilot in overcoming air loads.

- Figure 29 -

[3] Anti-servo Tab:

In figure 25 we see a stabilator having an anti-servo tab. Like the servo tab, this moves when the main surface moves, and the anti-servo action is obtained by the location of the anchor point for the actuator. Figure 30 shows the location of the jackscrew to get this motion. Both servo and anti-servo tabs are controlled from the cockpit so the pilot can trim the airplane for hands-off flight at any airspeed.

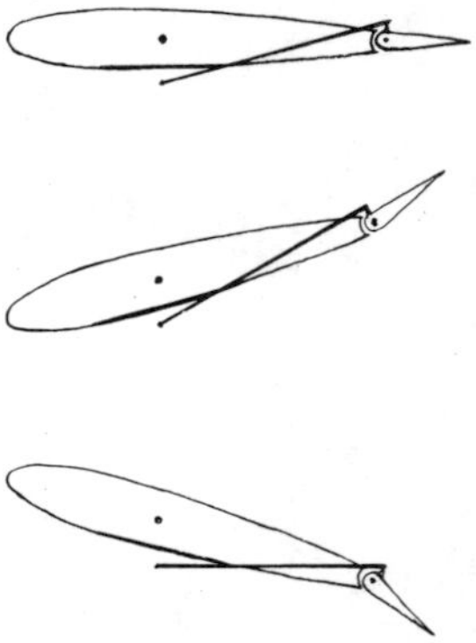

Anti-servo tabs aid in making the movement of the primary control less sensitive and prevent control flutter.

- Figure 30 -

[4] Adjustable stablizer:

Rather than use a tab to change the aerodynamic shape of the movable control surface, some aircraft pivot the stabilizer at its rear spar and raise or lower the leading edge with a jackscrew* to change the angle of incidence of the stabilizer, providing an adjustable tail load.

QUESTIONS:

36. Which aileron has the most travel, the one moving up, or the one moving down?
37. Does an anti-servo tab move in the same direction as the control to which it is attached, or in the opposite direction?
38. What is the difference between a simple flap and a Fowler flap?
39. Why do some aircraft have the leading edge of the stabilizer moved up and down by jackscrews?

D. FUEL SYSTEMS:

In a recent study conducted by the National Transportation Safety Board, it was reported that the greatest number of aircraft accidents attributable to mechanical malfunctions were the fault of the fuel system, or some component of this system. The problem lies in the complete lack of tolerance for any failure to provide proper fuel flow.

1. Fuel system types:

a. Gravity feed system:

The most simple system is that using gravity feed, in which the fuel tanks are located in the wings of high wing airplanes, and the fuel feeds by gravity through a filter into the carburetor. This system can be used only with engines having float type carburetors. Any gravity fuel system having a "both" position must have the tanks vented together with a common overboard vent.

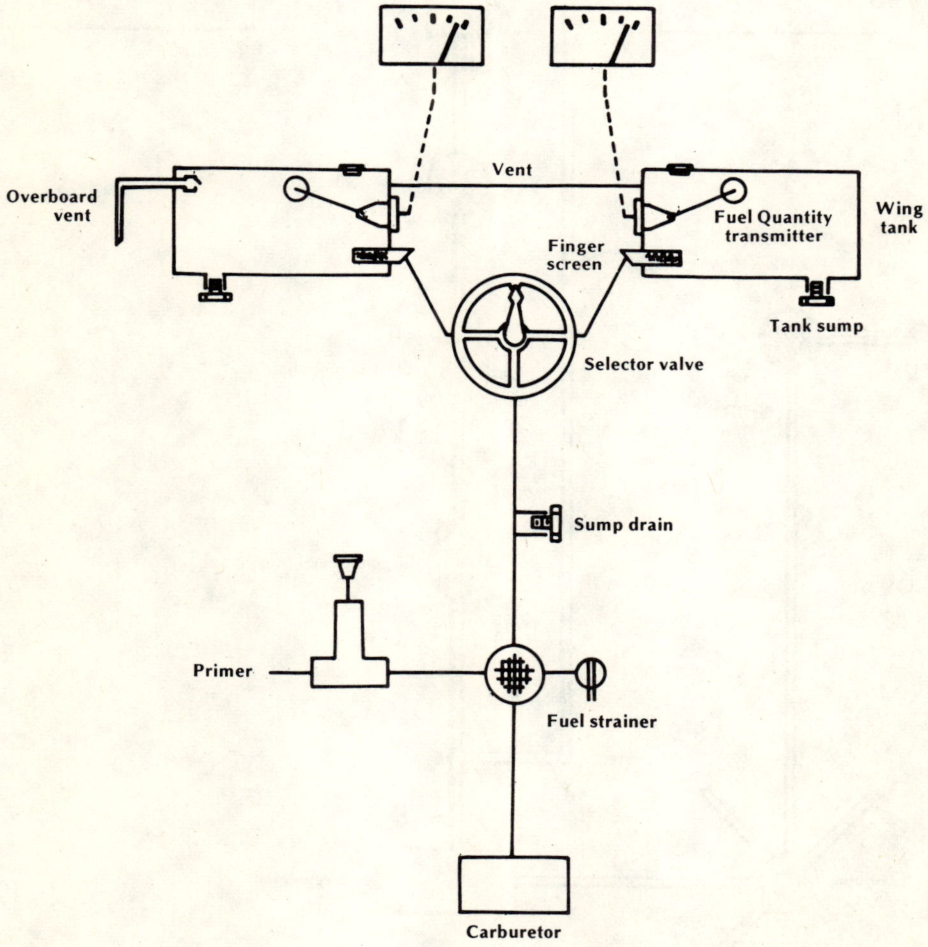

Simple gravity-feed fuel system of a high wing Cessna aircraft.

- Figure 31 -

b. Pump fed systems:

Low wing airplanes or airplanes having pressure carburetors or fuel injection systems must have a fuel pump to feed fuel from the tank into the engine. Pump fed systems cannot feed from both tanks at the same time because of the danger of emptying one tank and drawing air, causing the engine to stop when there may still be fuel in one tank. To have a simple system with only an "On" and "Off" position of the fuel valve, both wing tanks must feed into a common header or accumulator tank. A vent line connects the header and both main tanks, so actually this becomes one tank instead of two.

Any time the engine is fed by a pump, there must be one engine-driven and one auxiliary pump. These auxiliary pumps may be either hand operated or electrically driven.

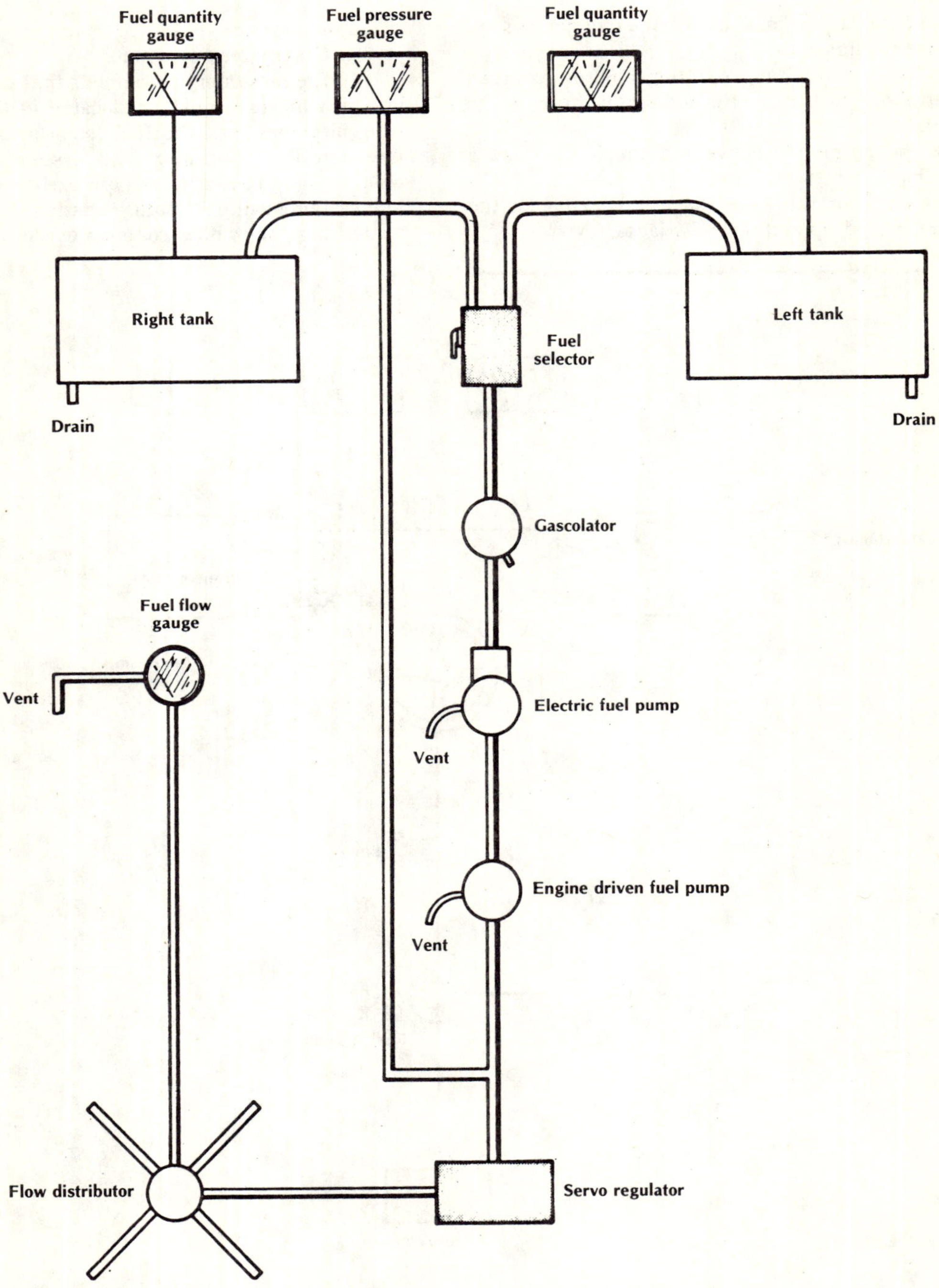

Pump-fed fuel system of a low wing Piper aircraft having a fuel injected engine.

- Figure 32 -

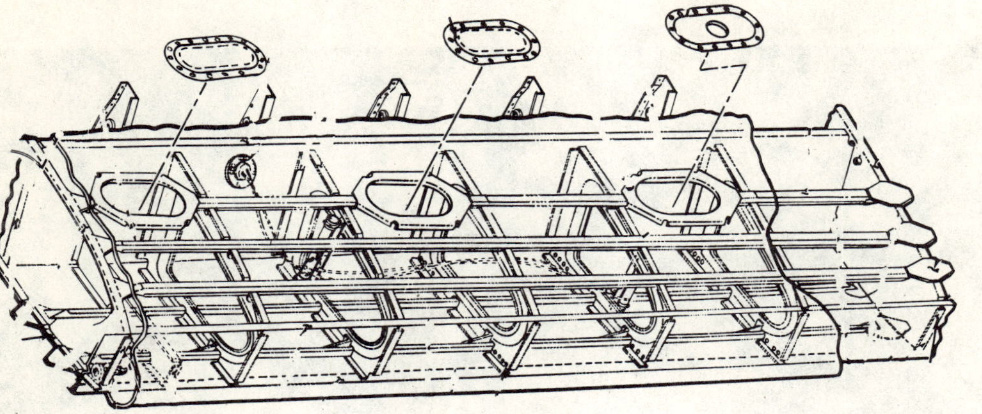

Cessna

An integral fuel tank is a portion of the wing structure sealed off and used to store fuel.

- Figure 33 -

2. Fuel tanks:

The quest for getting the most fuel into the available space has brought about the increased use of both bladder type fuel cells and integral tanks. A few airplanes, however, still use built-up tanks, welded of aluminum alloy and installed in a cradle mount in the wing.

a. Integral tanks:

One of the more popular tanks today is simply a section of the wing sealed off and used as a fuel tank - all of the structural seams sealed with a special compound. This type of tank has the advantage of light weight, as no separate tank need be carried, and of the greatest volume for the available space. On the negative side, it is possible for seep type leaks to develop around any rivets which have not been properly sealed.

b. Bladder tanks*:

Thin, neoprene impregnated cloth bladders are used in some airplanes. These bladders are laced or snapped into the tank cavities in the wing and have the fuel lines, vents, and fillers attached into them from their mounts in the structure. Tanks of this type are less likely to leak than integral tanks and are lighter-weight than a built-up metal tank would be.

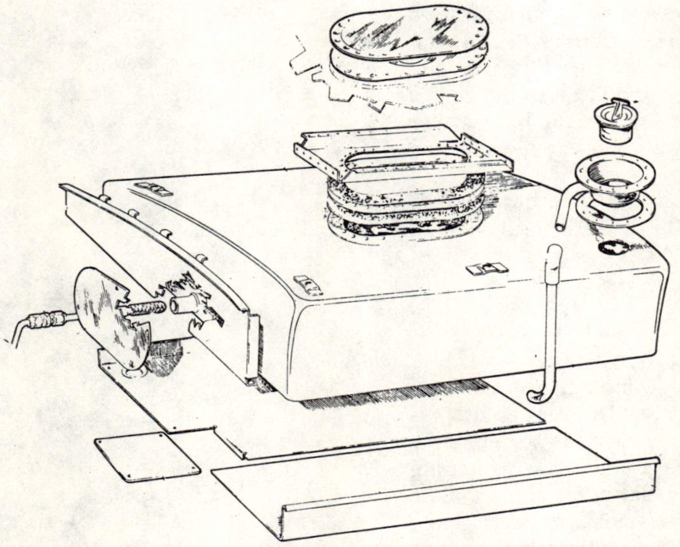

Piper

Bladder tanks are made of neoprene impregnated cloth, snapped or laced into the fuel cell cavity in the airplane wing.

- Figure 34 -

Bulkhead or positioning-type fittings are sealed with O-rings forced into a chamfer in the casting by a nut.

- Figure 35 -

3. Lines and Fittings:

Fuel tanks are connected to the carburetor by both flexible and rigid lines. The rigid lines are almost all made of annealed aluminum alloy. Flexible lines must be used between the fuel strainer and the carburetor to allow for vibration. These lines are most generally medium pressure hose similar to that used in the hydraulic system. This is not because of the pressure in the fuel system, but because of its high mechanical strength and its ability to withstand abuse.

Fittings are almost all standard AN hardware.* Bulkhead fittings sealed with O-rings are used in most strainers and carburetors.

4. Filters:

Fuel must not carry any contaminants into the carburetor or fuel injection system, and a series of filters is installed to prevent this. A fairly coarse mesh screen in the tank at the fuel outlet prevents large contaminants getting into the fuel lines. At the lowest point in the system is the main fuel strainer. A paper element is used quite successfully in some of these filters. Accordion pleated porous paper traps any contaminants which may be in the fuel. The paper passes the fuel, but stops any water that may be present. The most commonly used fuel strainer element for light aircraft is the fine wire mesh. Fuel can easily pass through this screen, but water or any other

Fuel strainers should be drained as part of a preflight inspection.

- Figure 36 -

contamination is held in the bottom of the bowl where it can be drained through the quick-drain valve on the pre-flight inspection.

5. Pumps:

a. Engine-driven pump:

Large aircraft engines normally use a vane type engine-driven pump to provide regulated fuel pressure to the carburetor. Engines equipped with the Continental fuel injection systems use a special vane type pump which is a component of the injection system. Turbo-charged engines having either fuel injection or pressure carburetors require the inlet fuel pressure to be maintained a specific amount above the inlet air pressure, and the engine driven pump is compensated to automatically increase the fuel pressure when the carburetor inlet air pressure is increased.

Engines requiring less fuel to flow often use a diaphragm type pump, similar to that used on automobile engines. The diaphragm is pulled down by a cam in the engine to draw fuel into the pump, and a calibrated spring forces the fuel into the carburetor. Pressure regulation is not nearly so precise with this type of pump as with the vane type.

Vane-type pumps such as this one used with the Teledyne-Continental fuel injection system are capable of providing precisely regulated fuel pressure.

- Figure 37 -

b. Auxiliary pumps:

Any fuel system requiring an engine-driven pump is required also to have an auxiliary pump. These may be mounted in the fuel tank, not only to provide fuel pressure for starting and pressure in case of engine pump failure, but also to pressurize the fuel in the line to prevent vapor formation at high altitude. Tank mounted pumps are usually of the centrifugal type, in which fuel is forced into the line by an impeller. Pumps of this type require no relief valve.

The plunger type pump, figure 38, uses an electric solenoid* to pull the fuel into the pump and a calibrated spring to force the fuel out into the system.

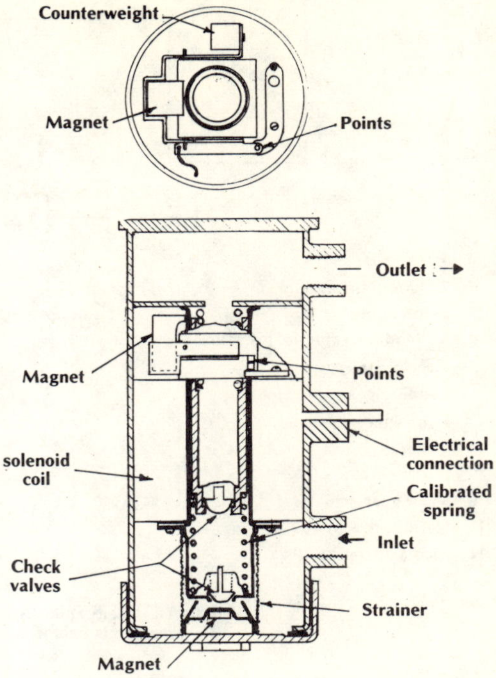

The calibrated spring forces fuel from the pump to the carburetor. The plunger is returned by an electromagnet.

- Figure 38 -

Fuel systems using a centrifugal type pump in the tank and a vane pump on the engine have these pumps in series so that all of the fuel flowing to the engine must pass through both pumps. Systems using a plunger type auxiliary pump and a diaphragm type engine pump have these pumps in parallel. The fuel flows through either of these pumps, but not both.

QUESTIONS:

40. Why are pump-fed systems not allowed to draw fuel from more than one tank at a time?
41. What is meant by an integral fuel tank?
42. How are fuel vapors prevented from forming in the fuel lines of aircraft flying at high altitudes?

E. HYDRAULIC SYSTEMS:

1. Types of systems:

Airplanes which have some of their maintenance performed by the pilot usually have very simple hydraulic systems. Normally the landing gear and flaps are all that are actuated by the system, as the brakes are independent.

There are two types of systems used: those using an engine-driven pump, and those having the pump operated by an electric motor.

a. Engine-driven pump systems:

Since the hydraulic pump is splined into the engine and turns all the time the engine is running, some provision must be made to unload the pump or prevent its having to work into system pressure all of the time. Two types of systems do this: the closed center system uses an unloading valve; and the open center system unloads the pump by circulating the fluid through the center of the selector valves.

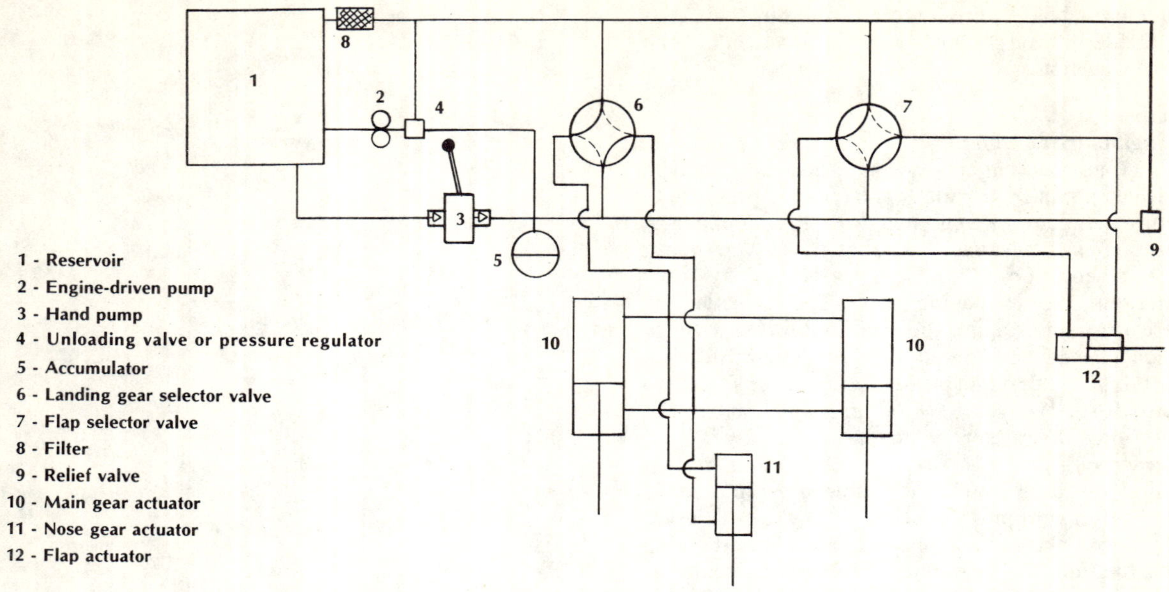

In a closed center hydraulic system, the selector valves are in parallel, and pressure is held in the system by an accumulator.

- Figure 39 -

[1] Closed center system:

Figure 39 shows a typical closed center system. Fluid is held in the reservoir which acts not only as a source of supply, but provides an expansion space for the fluid and allows entrapped air to escape. Fluid is drawn from the tank by the engine-driven pump which moves it through the system. To operate the system when the engine is not running, or in case of engine pump failure, a hand pump is provided. You will notice that the hand pump feeds from a point below the feed for the engine pump. In case a leak allows the engine pump to empty the reservoir, the hand pump still has enough fluid available to lower the landing gear. The unloading valve directs fluid from the pump into the system until a certain cut-out pressure is reached; then it shuts off the system and circulates the fluid back into the reservoir until the system pressure drops to a lower pre-determined pressure. Hydraulic fluid, since it is incompressible, cannot store pressure; so an accumulator* containing compressed air holds fluid in the system under pressure. A relief valve set at a pressure higher than the regulator prevents damage to the system in the event of regulator failure. A filter in the return line to the reservoir filters all fluid that has gone through the system. The selector valves direct the fluid to one side or the other of the actuator piston.

[2] Open center system:

The selector valves of the open center system, as shown in figure 40, are in series.

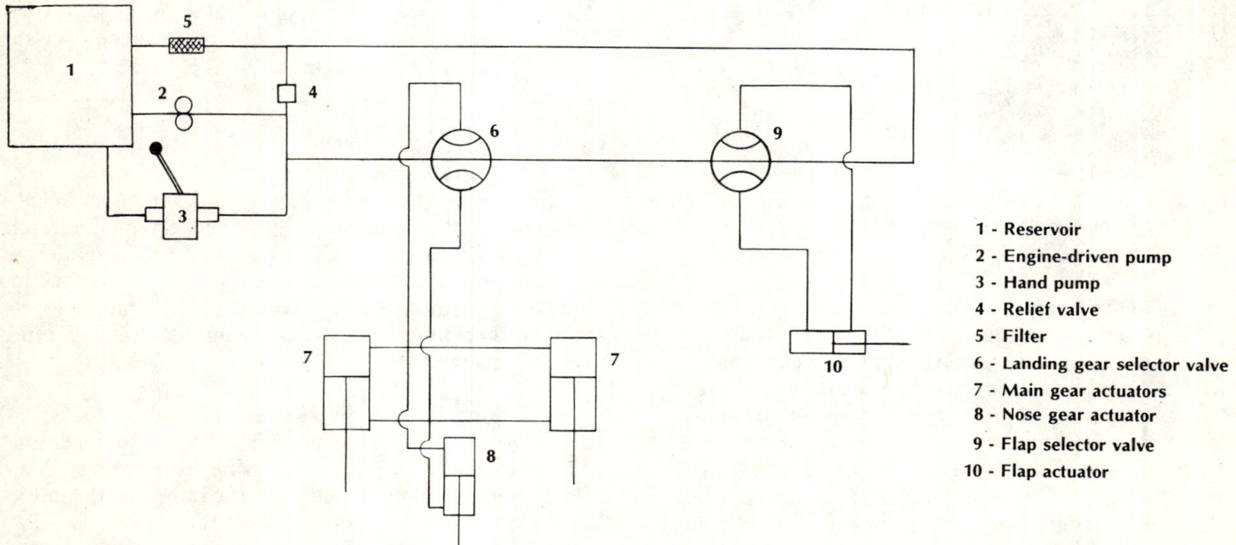

An open center hydraulic system has all of the selector valves in series and requires no accumulator. The selector valves serve as the unloading valve.

- Figure 40 -

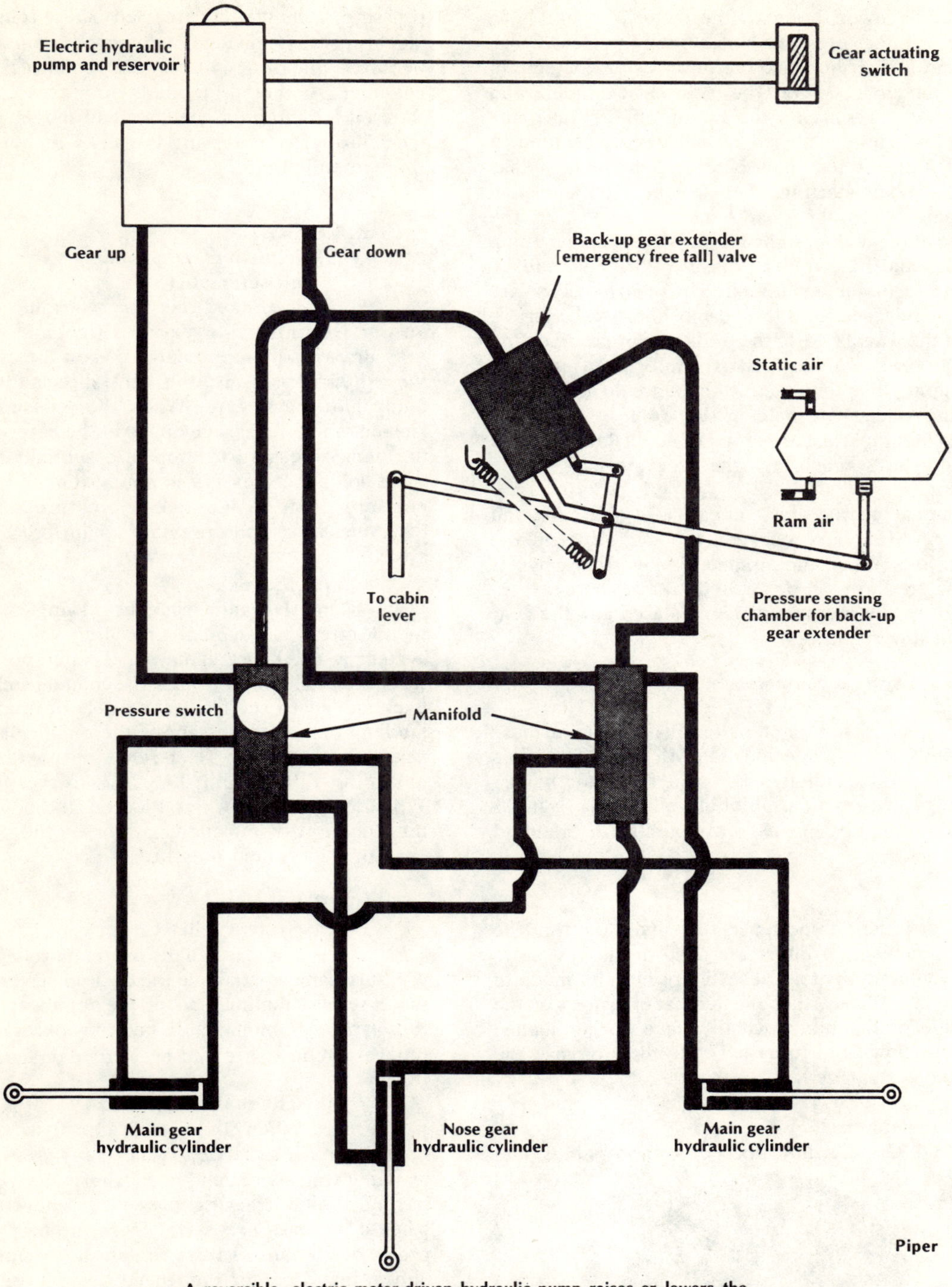

A reversible, electric motor-driven hydraulic pump raises or lowers the landing gear. A bypass valve is automatically held open at low airspeeds to prevent premature retraction of the gear on takeoff, and lowers the gear automatically when the airspeed drops off for landing.

- Figure 41 -

When no unit is being actuated, all of the fluid flows from the reservoir through the engine-driven pump, both selector valves, the filter, and back into the tank. When the gear selector is placed in the "Gear-Down" position, fluid is directed to the down side of the actuators. As the piston moves, fluid is forced out of the other side of the cylinder, through the selector valve, on through the flap valve, back to the reservoir. When all three cylinders have actuated and the pressure builds up, the gear valve automatically returns to its neutral position. This system needs no accumulator or unloading valve. A relief valve protects against high pressures until the selector valve returns to neutral after the system has actuated.

b. Electric pump systems:

Small aircraft with their limited requirements for hydraulic pressure often use electric motor-driven pumps. A typical system of this type is shown in figure 41.

Here we have an electric motor-driven pump located in the power pack. To lower the gear, the motor is operated in such a direction that pressure builds up in the Gear-Down manifold and is distributed to the down extension side of the three actuators. When the gear is extended and locked, a micro-switch cuts the motor off. To raise the gear, the gear handle is placed in the Gear-Up position. A safety switch on the left main landing gear torsion links prevents the motor starting when weight is on the main landing gear. When the gear handle is moved to the Gear-Up position with weight off the wheels, the pump motor starts, and fluid under pressure flows into the up side of the actuator cylinders. The down locks are hydraulically released, and the gear is allowed to retract. When all of the wheels are in the wells, the pressure on the Gear-Up line rises to about 1400 psi and opens a pressure switch, stopping the pump motor. There are no mechanical up-locks; the wheels are held in the wells by hydraulic pressure. When the pressure drops to about 1100 psi, the pump motor starts and brings the pressure back to the cut-out point.

This system incorporates, as a safety feature, an airspeed controlled valve between the Gear-Up and the Gear-Down line. When the airspeed is low, this valve is open and no pressure can be applied to raise the gear; or, if the gear is up, the pressure will be released and the gear will free-fall down and lock.

2. Hydraulic system components:

a. Fluid:

Almost all piston engine aircraft use mineral base hydraulic fluid meeting specification MIL-H-5606. This is red colored and is essentially a kerosene type fluid having anti-foaming and corrosion inhibiting additives. Systems using this fluid may be flushed with naphtha or Stoddard solvent.

b. Filters:

It is of utmost importance that hydraulic fluid be kept perfectly clean, so filters are used in the system to remove any contaminants. These filters may be made of fine wire mesh or of accordion pleated paper. Filters on the pressure side of the pump usually have spring loaded valves which allow fluid to by-pass the filter in case the element becomes clogged.

c. Accumulators:

Since hydraulic fluid is non-compressible, and some provision must be made for storing it under pressure, accumulators are used. These consist usually of steel spheres, divided into two compartments by a neoprene diaphragm or bladder. Compressed air fills one chamber. As the pump builds up pressure in the system, fluid enters the fluid chamber and moves the diaphragm over, further compressing the air. This holds pressure on the hydraulic fluid.

An accumulator holds hydraulic fluid under pressure from compressed air or nitrogen.

- Figure 42 -

d. Valves:

[1] Regulators or unloading valves:

[a] Automatic:

Closed center systems may use an automatic unloading valve which maintains system pressure with an accumulator. When the system is below the regulated pressure, this valve directs the output of the pump into the system. When the pressure rises to the kick-out point, a passage opens to the reservoir and allows the pump pressure to drop to essentially zero. A check valve holds the pressure in the system. When the system pressure drops to the kick-in pressure, the line to the reservoir closes and the pump again forces fluid into the system.

[b] Manual:

Hydraulic systems having simple requirements often use manual unloading valves. Any time a system is not being actuated, the pump circulates fluid from the reservoir, through the manual unloading valve, back into the reservoir. When it is desired to lower the landing gear, for instance, the pilot places the gear selector valve in the Gear-Down position and closes the unloading valve. Fluid then goes to the gear actuators. When the gear is down and locked, the pilot then returns the valve to the unloaded position so the pump does not have to hold system pressure.

[2] Relief valve:

[a] System relief valve:

There is usually one main relief valve set at a pressure somewhat above the desired system pressure. In the case of a malfunction of the regulator, or if for any reason the system pressure becomes excessive, this valve relieves the pressure and prevents damage.

[b] Thermal relief valve

When fluid is trapped in a line, some provision must be made to relieve any pressure generated by fluid expansion due to heat. Thermal expansion valves, having a high cracking pressure, relieve this pressure when it becomes excessive. Since the part of the system protected by these valves contain no compressible fluid, only a few drops of fluid need be relieved to lower the pressure to an acceptable level.

[3] Sequence valves:

Some landing gear systems use hydraulically operated wheel well doors. One way of being sure the doors open fully before the gear extends is to employ a sequence valve. Only when the door opens fully and actuates the sequence valve can fluid be directed into the gear extension cylinder.

[4] Priority valves:

Another type of valve which allows sequential operation of the gear doors and actuators is the priority valve. Rather than requiring mechanical contact with the actuated gear door, this valve is pressure actuated. Doors require less pressure than the gear actuators and the priority valve in series with the gear actuator remains closed until the wheel well doors open completely, and pressure builds up against them. The pressure will then rise high enough to open the priority valve and move the gear.

QUESTIONS:

43. What are two ways aircraft hydraulic pumps may be driven?
44. Why does the engine-driven hydraulic pump not feed from the bottom of the reservoir?
45. What is the function of a pump unloading valve?
46. What is the function of an accumulator?
47. What color is mineral base hydraulic fluid?
48. What is the purpose of a relief valve?
49. What is the function of a priority valve?

F. ELECTRICAL SYSTEM:

Without question, the most used and most taken for granted system in an airplane is the electrical system. We use it to start the engine, operate the instruments, power the radio, light the lights, and on some airplanes operate the flaps and landing gear. Each year, electrical loads go up as new and more sophisticated equipment is added to the list available to private aviation.

1. Generating systems:

There are two types of generating systems you may find on your aircraft: those having a generator* and those having an alternator*.

a. Generator systems:

The generator which has been used for years as the main source of direct current electricity (DC), figure 43, actually manufactures alternating current (AC) in the rotating member called the armature.

The direct current generator is the main source of electrical energy for many light general aviation aircraft.

- Figure 43 -

The electrical energy is taken from the armature by carbon brushes riding on the copper bars of the commutator*. The commutator is actually a mechanical switch to convert the AC into DC and allow only relatively pure direct current to enter the aircraft electrical system.

The amount of voltage produced by a generator is determined by the amount of current allowed to flow in the field coils, figure 44.

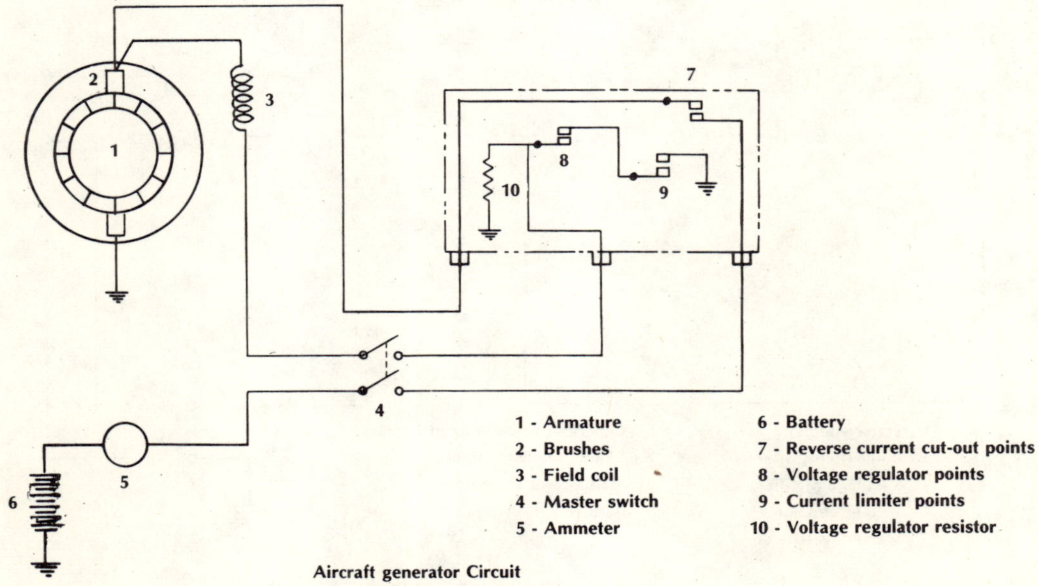

Aircraft generator Circuit

- Figure 44 -

When the master switch is turned on and the engine started, the generator begins to turn and generate a voltage. This causes some current to flow through the field coil and through the voltage regulator and current limiter points to ground. This field current causes more voltage to be generated, and the output rises to a value somewhat above that of the battery. When this voltage is sufficiently high, the reverse current cut-out contacts close and connect the generator to the battery circuit. The voltage continues to rise and the field current increases until the desired voltage is reached. The voltage regulator points then open and decrease the field current, dropping the output voltage. These points vibrate continually maintaining the voltage at the required level. If, for any reason, an excessive load should be placed on the generator, the current limiter points, also in series with the field coil, open. The voltage drops low enough that excess current can no longer flow.

A reverse current cut-out is installed between the generator and the battery to prevent the battery discharging through the generator. As soon as the generator voltage increases above the battery voltage, the relay closes and connects the generator to the battery. When the generator voltage drops below the battery voltage, there will be a momentary reverse flow of current which causes the contacts to open and disconnect the generator from the battery.

b. Alternator System:

Semi-conductor diodes* are now available which are not seriously affected by heat, so alternating current produced in alternators may be converted into direct current without the use of a commutator. The generated current may be produced in the fixed winding, or the stator*, and not have to be picked up and switched by brushes. Alternators are short and somewhat large in diameter, to allow for eight sets of magnetic poles for the field and three sets of windings in the stator.

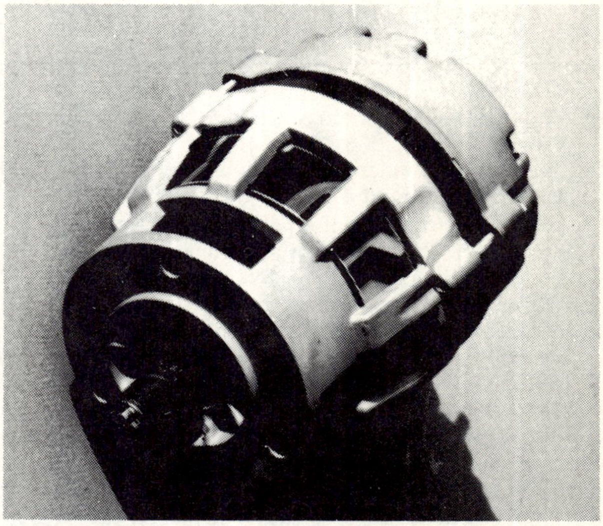

Aircraft alternators can produce their rated output at low RPM because of a multi-polar rotating magnetic field.

- Figure 45 -

The full output current can therefore be produced at a much reduced RPM.

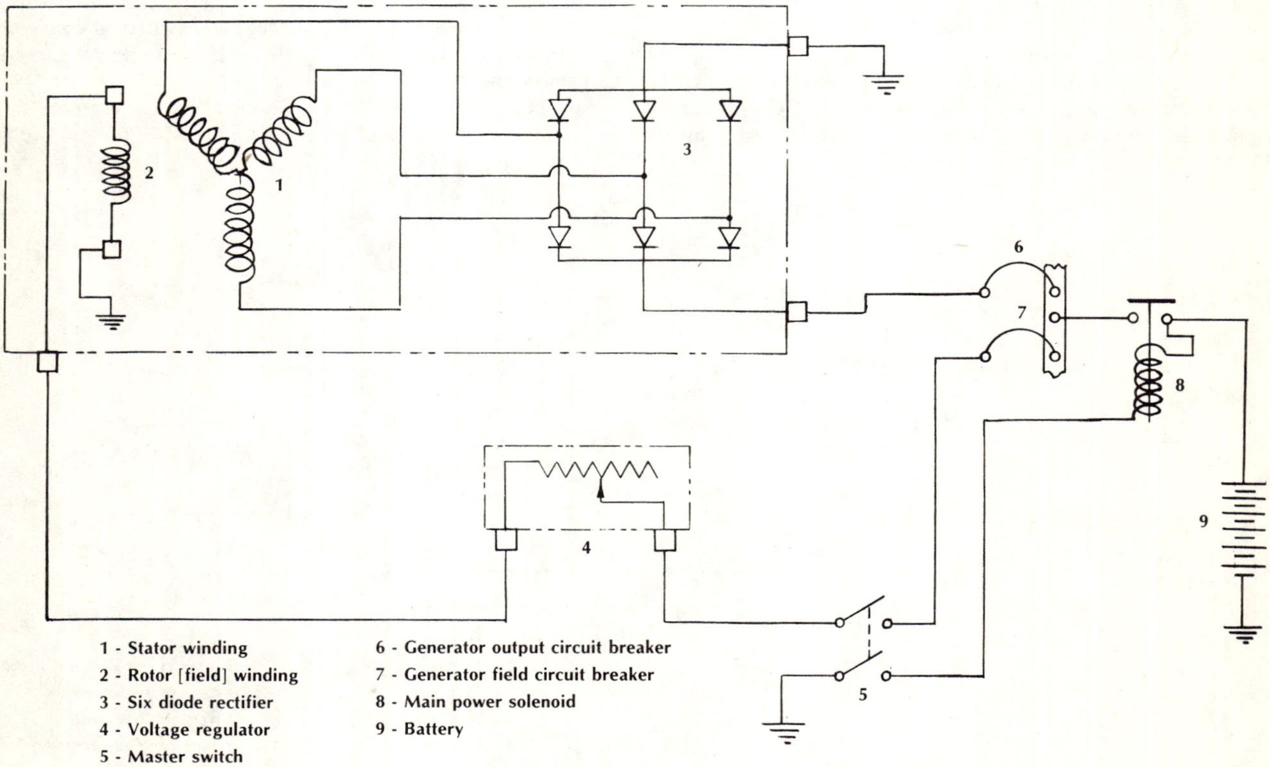

1 - Stator winding
2 - Rotor [field] winding
3 - Six diode rectifier
4 - Voltage regulator
5 - Master switch
6 - Generator output circuit breaker
7 - Generator field circuit breaker
8 - Main power solenoid
9 - Battery

Alternator Circuit

- Figure 46 -

The rectifying diodes do not allow current from the battery to flow into the alternator windings, so there is no need for a reverse current cut-out. The field is not supplied with current directly from the alternator, so field current may be limited to prevent the alternator exceeding its current rating. Therefore, there is no need for a current limiter. The voltage regulator may be of either the vibrating points or transistor type, either of which acts simply as a variable resistor between the positive battery bus* and the alternator field. When the voltage drops, the effective resistance decreases, allowing more field current to flow, and brings the output voltage back up to the desired value.

2. Wiring and Circuit Protection devices:

a. Wire:

Almost all of the wire used for lights, radio, and control circuits is stranded copper, covered with a plastic insulation. Any wire installed in an airplane must be of sufficient size that there will not be too much heat generated from the current carried, or an excessive voltage drop caused by the resistance of the wire. Before any electrical equipment is installed by an A&P, he must determine that these two criteria are met by the wire he uses.

Any wire must be shielded which carries alternating current. This means the wire is encased in a tinned copper braid and the braid grounded. Any electromagnetic waves which radiate from the wire are picked up by the shielding and carried to ground. The shielded wire most familiar to pilots is that in the spark plug leads.

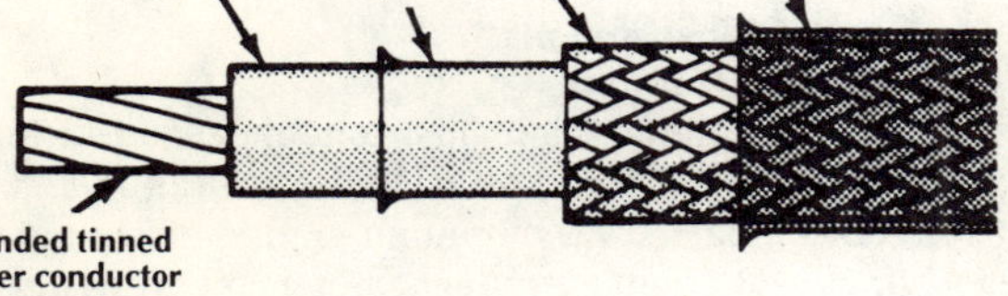

The tinned copper braid intercepts electromagnetic radiation and conducts it to ground, preventing radio interference.

- Figure 47 -

b. Circuit protection devices:

[1] Fuses:

A short circuit may easily pass so much current through a circuit that the heat could cause a fire. To prevent this, a fuse is installed between the electrical bus and the load. In case the current exceeds the rating of the fuse, this metal link will melt and open the circuit.

[2] Circuit breaker:

Rather than requiring a pilot to fumble around to replace a fuse during a critical time in flight, most manufacturers now use circuit breakers in all circuits which may be restored in flight. If a fault occurs, the excessive current produces either enough heat or a strong enough magnetic field to open a set of switch contacts and stop the flow.

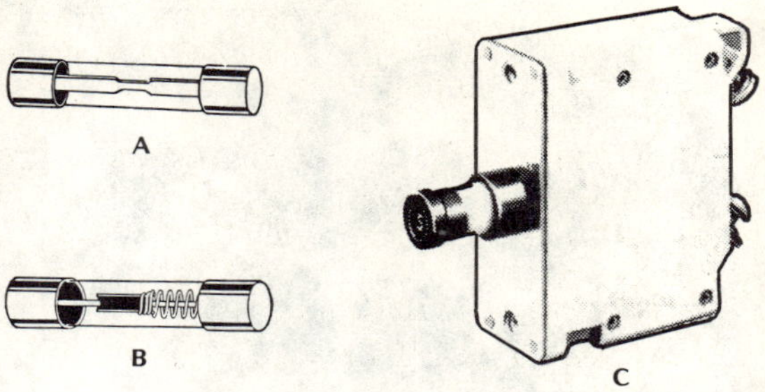

A - Regular tubular fuse; B - Slow blow fuse; C - Button-type circuit breaker

Circuit Protection Devices

- Figure 48 -

Sometimes circuit breakers may pop out because of transient currents in the circuit instead of a true fault. Pushing the button back will restore the circuit. If a fault actually exists, the circuit will immediately re-open.

3. Batteries:

The primary source of electrical power in an airplane is the generator or alternator, but the battery is a source of stored electrical energy.

Aircraft batteries serve as a reservoir for electrical energy when the generator is not operating, and as a voltage stabilizer when the generator is carrying the electrical load.

- Figure 49 -

There are two types of batteries used in aircraft, the old and faithful lead-acid battery, and the newer, high output nickel-cadmium batteries. Since almost all airplanes subject to pilot-maintenance will use lead-acid batteries, these will be the ones discussed here.

Batteries do not produce electricity, but, rather, store it in the form of chemical energy, ready to convert it back into electrical energy when the circuit to some unit is completed. Chemical energy exists when the plates within the battery contain different compounds of lead. An electrolyte of dilute sulfuric acid covers the plates, and, as the battery is discharged, both plates convert to lead sulfate. When the generator voltage is higher than that of

Cessna

The instrument panel of a typical, well equipped, single-engine personal airplane.

- Figure 50 -

the battery, electrical energy is put back into the battery and the plates change their chemical composition. Hydrogen gas is liberated from the electrolyte during this charging process.

Almost all of the light aircraft use 12 volt batteries, similar to but smaller than automobile batteries, and therefore cannot store as much electrical energy. For this reason, it is important that auxiliary power units or booster batteries be used when engine starting is difficult, as in cold winter weather when the oil is thick and the efficiency of the battery has decreased.

QUESTIONS:

50. What is the purpose of the commutator and brushes in an aircraft generator?
51. How is the output voltage of an aircraft generator controlled?
52. What component in an alternator takes the place of the commutator in a generator?
53. What two things must be considered when selecting the wire to use in an aircraft electrical installation?
54. What is the purpose of shielded wire?
55. What is used as the electrolyte in a lead-acid battery?

G. INSTRUMENT SYSTEMS:

There is no question but that the increased utility of the airplane has come about chiefly because of reliable instruments. Flying has been made both more safe and more efficient since we have been freed from the necessity of flying by pilotage beneath the overcast.

1. Required instruments:

a. Day VFR flight:

The minimum flight instruments required for this type of flying are: airspeed indicator, altimeter, and magnetic compass. The minimum engine instruments for each air cooled engine are: tachometer, oil pressure gage, oil temperature gage, and a fuel quantity gage for each tank. A manifold pressure gage is also required if it is possible to supercharge the engine to a manifold pressure above that for which the engine is rated. If the airplane has a retractable landing gear, an indicating system must show the position of the gear.

b. Night VFR flight:

In addition to all of the instruments required for day VFR flight, the aircraft must have an adequate electrical system, complete with a spare set of fuses; a red, a green, and a white position light; and an approved anti-collision light.

c. Instrument flight rules:

All of the equipment specified for day or night VFR must be installed for IFR flying plus a two-way radio communication system and navigation equipment appropriate for the ground facilities used; a turn and slip indicator or turn coordinator, a sensitive altimeter, a clock with a sweep second hand, a gyro horizon, and a directional gyro. A generator adequate to carry all of the electrical load is also required.

2. Engine Instruments:

a. Tachometers:

Two types of tachometers are in common use in light aircraft, mechanical and electrical. The mechanical is driven by a flexible shaft from the engine. This type normally incorporates an hour meter to show the hours the engine has operated. It is geared directly to the engine and is accurate only at cruise RPM. The slight error caused by operation above or below this speed is of little real concern.

Tachometers for twin engine aircraft are usually electrically operated because of the long distance between the engine and the instrument. One of the more popular electric tachometers takes its signal from a small coil mounted in one of the magnetos. The AC voltage generated in this coil by the magneto magnet is converted to DC within the instrument and moves the pointer over the dial to indicate the RPM the engine is turning.

b. Oil temperature indicator:

These are usually pressure gages which sense the pressure of a vapor in an enclosed tube. A bulb containing a highly volatile liquid is inserted into the oil system, usually in the engine oil strainer. This is connected to the instrument by a continuous capillary tube. As the oil temperature increases, the pressure of the vapors above the liquid increases and causes the instrument to read higher. The dial is calibrated in degrees of temperature rather than pressure.

c. Cylinder head temperature:

The higher temperature of the cylinder head requires an electrical instrument for its indication. There are two types in use, one working only when the electrical master switch is turned on.This is a resistance change instrument, and uses a probe having about one hundred ohms resistance at normal room temperature. A spring in a special adapter holds the bayonet probe* tight against the cylinder head. As the head gets hotter, the resistance of the wire increases and the indication goes higher.

The other type of cylinder head temperature indicator uses a thermocouple, a special lead consisting of two different types of wire. A voltage is generated in these wires which is proportional to the temperature difference between the end of the lead at the engine and the end in the instrument. Since the thermocouple generates its own electricity, it will operate whether or not the electrical master switch is on. The probe for this indicator may be either a bayonet type, or one whose leads are imbedded in a special spark plug gasket.

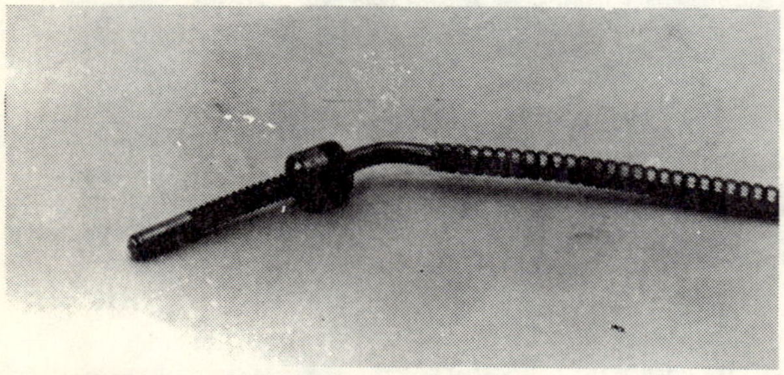

The bayonet-type cylinder head temperature probe is tip-sensitive and is held in tight contact with the head by a spring.

- Figure 51 -

d. Manifold pressure gage:

The two instruments which tell a pilot the power his engine is developing are the tachometer and the manifold pressure gage. While manifold pressure is not a direct indication of the pressure inside a cylinder, it does relate to this pressure. The more absolute pressure* there is in the intake manifold, the more oxygen there will be in the cylinder, and the more energy will be released from the fuel. A manifold pressure gage uses an evacuated and sealed bellows to measure absolute pressure; that is, pressure above a vacuum. The gage connects into the intake manifold, somewhere between the carburetor and the intake valves in the engine.

The manifold pressure gage measures the absolute pressure in the intake manifold of the engine.

- Figure 52 -

When the engine is not running, it measures the pressure of the atmosphere surrounding it. When the engine is idling, it is trying to pump more air than the carburetor allows to pass, so the pressure downstream of the carburetor is less than atmospheric. A non-supercharged engine can never have a manifold pressure greater than atmospheric since this is the only pressure pushing air into the cylinders. Turbocharged engines, on the other hand, can have a pressure greater than atmospheric, and it is with this type of engine that manifold pressure becomes a vital concern. If the manifold pressure becomes excessive, more heat energy will be released than the engine can absorb, and detonation with severe damage may result.

e. Fuel pressure gage:

A bourdon tube* pressure gage is used to indicate the fuel pressure at the inlet of the metering system. To use one gage to check both the engine-driven and auxiliary pumps, turn on the auxiliary pump and note the pressure before starting the engine. Then, after the engine is running, turn the auxiliary pump off and read the pressure from the engine-driven pump.

f. Fuel flow indicator:

Normally, light aircraft do not have fuel flow meters which actually measure the flow of the fuel in the line. Large piston engines and almost all turbine aircraft use this type, but the gage in light aircraft whose dial reads "Fuel Flow" is actually a pressure gage, reading pressure drop across the fuel injector nozzles. Since the pressure drop across fixed orifices, such as these nozzles, is proportional to the amount of flow, these gages may legitimately be calibrated in units of flow.

g. Fuel Quantity:

It has been succinctly said that the only accurate way of telling the amount of fuel in a tank is to look into it. The fuel quantity indicating system on many of the smaller aircraft do actually leave something to be desired, as far as accuracy is concerned. The limited scale on the indicator dial is indicative of the accuracy we can expect from the system. There are two types of systems in general use today:

[1] The resistor type:

A float rides on the top of the fuel in the tank and moves the wiper arm of a variable resistor across the resistance unit. The change in this resistance causes a change in the current through the indicator. This arrangement allows the indication of the dial to relate to the position of the float in the tank.

A float rides on the top of the fuel and moves the arm of a variable resistor to measure the quantity of fuel in the tank.

- Figure 53 -

[2] Capacitance type:

A tubular probe having no moving parts is mounted in the fuel tank. This probe, which is actually an electrical capacitor, is connected into a bridge circuit which measures the electrical capacity of the probe. When the tank is empty, the insulator in the capacitor is air, with a constant of one. When the tank is full, the insulator is fuel having a constant of two, so that filling the tank causes the capacity of the probe to almost double. The indicator is in a bridge circuit, but, instead of reading in units of electrical capacity, reads in gallons or pounds of fuel.

h. Exhaust gas temperature gage:

A thermocouple probe is inserted into a hole drilled into the exhaust stack near the cylinder.

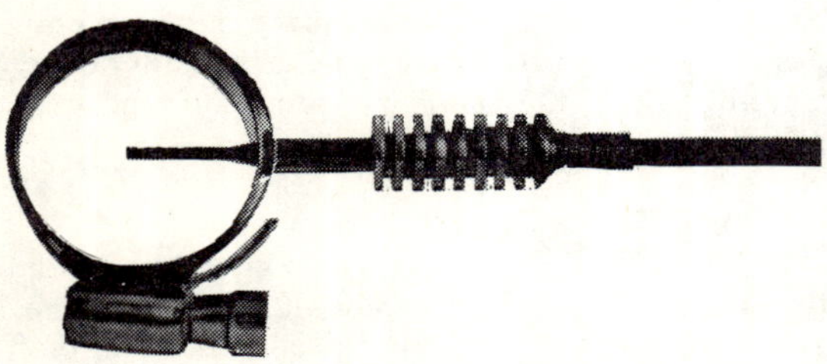

A thermocouple probe stuck into the exhaust pipe measures the temperature of the exhaust gas to indicate the mixture ratio of the fuel being burned.

- Figure 54 -

This senses the temperature of the exhaust gas and displays it on an instrument similar to a cylinder head temperature gage. An overly rich mixture has an excess of fuel, some of which lowers the temperature of the exhaust gas. An excessively lean mixture cools the exhaust gas with its excess air. A mixture that has exactly the proper amount of fuel for the air produces the hottest exhaust gas. This mixture is not desired for normal operation but is a dependable reference point for proper leaning. This instrument will be discussed further in the section on engine operation.

3. Flight instruments:

a. Altimeter:

No other instrument is so critical to the safety of flight as the altimeter. On instrument approaches its accuracy is vital. The pressure altimeter was one of the first flight instruments, and its principle of operation has remained basically unchanged since the late 1920's when the sensitive altimeter with its barometric adjustment was introduced.

The altimeter contains a sealed and evacuated bellows, a series of levers and arms, and a transmission mechanism to amplify the small movement of the diaphragms into a large movement of the pointer. When the barometric scale is adjusted to the existing sea level pressure, the pointers indicate the altitude of the airplane above that pressure level. Since all aeronautical charts show the heights of obstructions in feet above mean sea level, height above the obstruction or the terrain can be determined. The altimeter is the only non-electronic instrument to specifically require an accuracy check. Every twenty-four months all altimeters used for instrument flying must be checked for calibration and a record kept with the airplane.

The altimeter, along with the airspeed indicator and the rate of climb indicator, is connected into the static system. The pick-up for this system is usually a small hole in the side of the airplane, figure 55. The static system must be checked for leaks every twenty-four months if the airplane is operated under instrument flight rules.

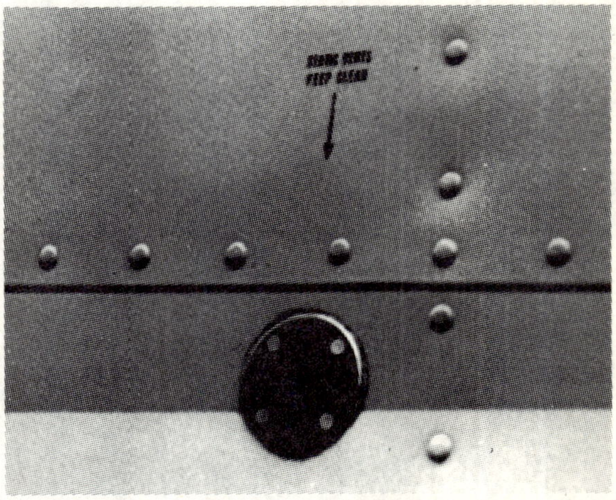

Static ports on the side of the airplane are the source of static, or still, air for the airspeed indicator, altimeter, and rate of climb indicator. They must be kept open at all times.

- Figure 55-

b. Airspeed indicator:

The lift produced by an aircraft wing is proportional to, among other things, dynamic pressure. This is a differential between static, or still air pressure and ram, or pitot, pressure. The faster the airplane moves through the air, the greater this differential becomes and the higher the indication of airspeed.

Ram air pressure is picked up by the pitot tube, which is an open ended tube, sticking directly into the air flow, usually on the wing, outside of the propeller slipstream.

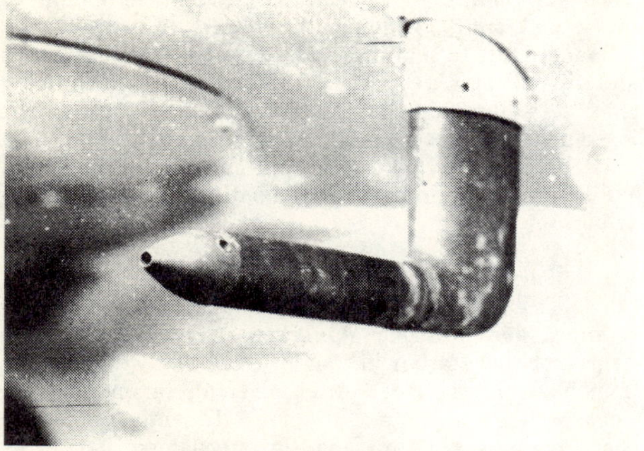

An open tube, facing directly into the airstream, serves as a pickup for pitot or ram pressure for the airspeed indicator.

- Figure 56 -

Static air comes from the same source used by the altimeter.

c. Rate of climb indicator:

The rate of climb, or vertical speed, indicator is a rate instrument which measures the rate of change in air pressure as the airplane climbs or descends. It consists of an airtight case enclosing a phosphor bronze capsule, connected to the static port. There is a very tiny calibrated leak between the capsule and the inside of the instrument case.

As the airplane climbs, the pressure inside the capsule decreases. The case pressure changes too, but because this change must come through the calibrated leak, it lags behind that in the capsule. The faster the change, the greater the lag, and the greater the indication on the dial. A decrease in altitude is indicated as the pressure increase in the case lags behind the increase inside the capsule.

The lag in the vertical speed indication makes this instrument unusable as a primary pitch indicator for instrument flight, but this lag is eliminated in the instantaneous vertical speed indicator. An accelerometer in the case starts the pressure change as the airplane pitches, and by the time the indication caused by the pitch has stabilized, the indication caused by the pressure change has taken effect. This results in an instantaneous indication of pitch change.

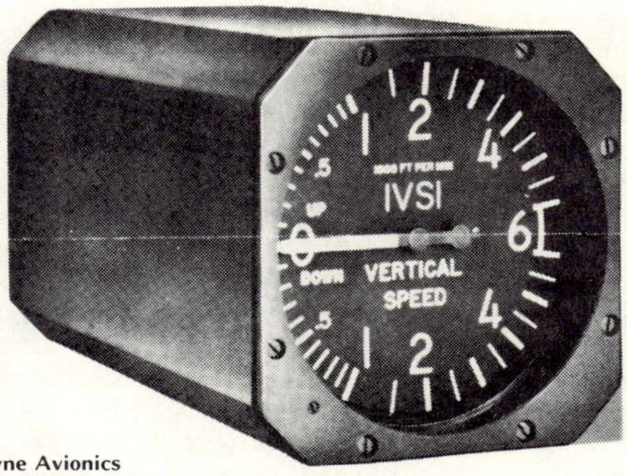

Teledyne Avionics

An accelerometer in the Instantaneous Vertical Speed Indicator overcomes the lag usually associated with ordinary rate of climb indicator.

- Figure 58 -

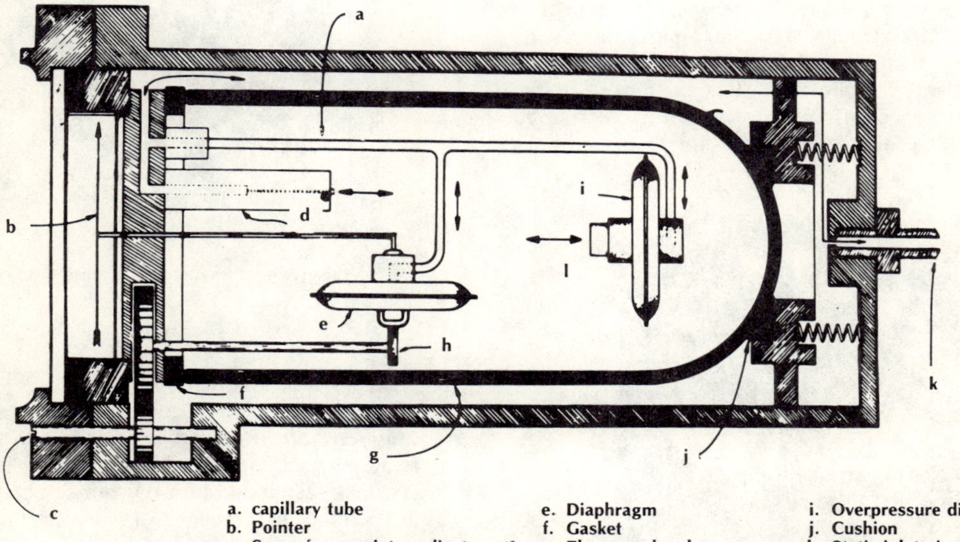

a. capillary tube
b. Pointer
c. Screw [zero pointer adjustment]
d. Diffuser valve
e. Diaphragm
f. Gasket
g. Thermos chamber
h. Cam
i. Overpressure diaphragm
j. Cushion
k. Static inlet nipple
l. Relief valve

The vertical speed, or rate of climb, indicator operates on the principle of a calibrated leak between the inside of the diaphragm and the case, causing the pointer to indicate the rate of pressure change.

- Figure 57 -

QUESTIONS:

56. What does the manifold pressure gage actually measure?
57. What does the fuel flow indicator of an injected engine actually measure?
58. What type pickup is used with an exhaust gas temperature indicator?
59. Name three instruments which connect to the aircraft static system.
60. What does an airspeed indicator actually measure?
61. What does a vertical speed indicator actually measure?

d. Turn and slip indicator:

The turn and slip indicator, needle and ball, or turn and bank indicator - regardless of what it is called - is one of the first and most dependable instruments used for flying when there is no outside reference. There are two instruments in one case; one, a simple curved glass tube, partially filled with a clear liquid and housing a black glass ball, about the size of a marble. When the airplane is level and not turning, the force of gravity alone acts on the ball, and it stays in the center of the tube. In a flat turn, centrifugal force moves the ball to the outside of the turn. If the airplane is banked correctly for the rate of yaw,* centrifugal force cancels the effect of gravity and the ball remains in the center of the curved tube. If the angle of bank is excessive, the ball rolls to the inside of the turn, and if it is insufficient, centrifugal force predominates and the ball goes to the outside of the turn. This ball, formerly called the bank indicator, as you see tells nothing about the amount of bank; so the name has, in recent years, been more correctly changed to slip indicator.

The other instrument in the case is the turn indicator, actuated by a small wheel suspended in a frame called a gimbal. The axis of the gimbal coincides with the longitudinal axis* of the airplane, and the spin axis aligns with the airplane's lateral axis.* The wheel is spun at a high speed by an electric motor or by an air stream. As the wheel spins, a force known as precession* acts on it at 90 degrees to the point of application. Since the gimbal is free to rotate about the longitudinal axis, yet is restrained about the vertical, a force is applied to the front and back of the gimbal as the airplane yaws (rotates about its vertical axis). This force is not felt at the front and back, but at the top and bottom of the wheel, causing it to lay over, or try to rotate about its gimbal axis. This rotation is restrained by a calibrated spring, and the gimbal moves a pointer across the dial to indicate the rate of yaw.

The turn indicator is not calibrated in number of degrees per second, but the wide pointer deflects its own width for a standard rate turn of three degrees per second. The turn and slip indicator is especially important in rough air flying since it is incapable of tumbling. If the limits of the instrument are exceeded, the pointer simply bears against its stops until the rate of yaw returns to within its limits. While formerly a primary instrument, it is now secondary or back-up.

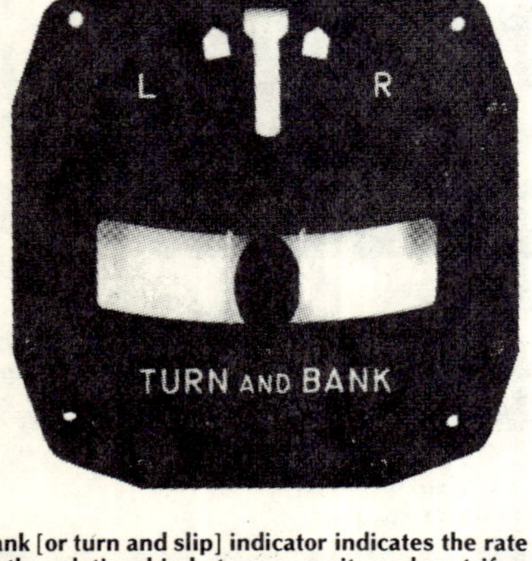

The turn and bank [or turn and slip] indicator indicates the rate of yaw with the needle, and the relationship between gravity and centrifugal force with the ball.

- Figure 59 -

e. Turn coordinator:

The turn and slip indicator measures the rate of rotation about the airplane's vertical or yaw axis, but since a turn actually starts with the lowering of a wing, a more immediate indication of a turn is given if rotation about the roll as well as the yaw axis is sensed. Essentially, the turn coordinator uses a turn and slip mechanism having its gimbal axis tilted about thirty degrees up from the longitudinal axis of the airplane. In this way, as you can see in Figure 60, rotation about the longitudinal (roll) axis, as well as yaw is sensed and the indicator provides an instantaneous indication of the turn, starting as soon as the wing drops and continuing all during the yaw. As the name implies, it aids the pilot in keeping the turn coordinated.

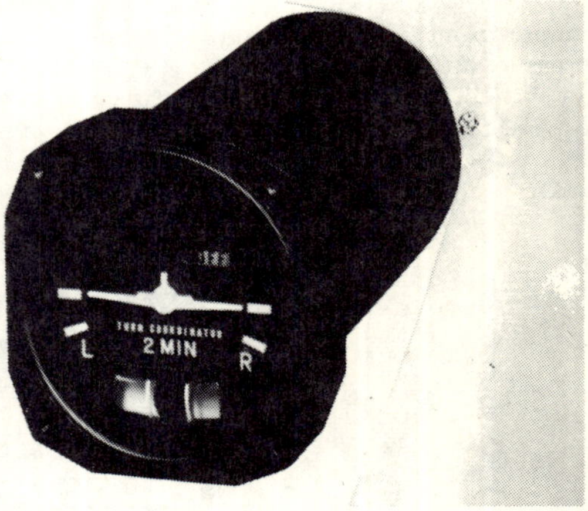

The gyro in a turn coordinator senses both roll and yaw.

- Figure 60 -

f. Magnetic compass:

One instrument required in all aircraft is some form of magnetic direction indicator. The most simple is the floating magnet compass. This consists simply of a small brass float suspended in a bowl of compass fluid, a clear liquid similar to kerosene. Pivots in the float keep it centered in the bowl, and two small permanent magnets soldered to the bottom of the float align with the earth's magnetic field. Around the float is a strip of metal called the card marked with the four cardinal headings and degree indications between the cardinals. A glass window in the bowl allows the pilot to see the card, and a vertical marker called the lubber line references the nose of the airplane to the card.

The magnets align with the earth's magnetic field and the number seen across the lubber line tells the pilot his heading relative to magnetic north. This system of direction indication has sufficed for centuries, but its limitations must be considered when it is used for flight navigation.

First, the compass points to magnetic north and all charts are oriented to geographic, or true north. This error, variation, must be lived with, and a correction applied by the pilot. Variation error is dependent upon the location of the airplane and does not change with its heading.

The compass magnet does not care whether it aligns with the earth's field or some other magnetic field. The airplane, its engine, and any wire carrying electric current are sources of magnetism, and the compass magnet is influenced by these fields. Since all fields concentrate about a point on the airplane, on some headings this concentration will increase the effect of the earth's field, and on others, it tends to cancel it. Deviation error, as this is called, is dependent upon the heading of the airplane and is not affected by the location. An A&P technician can compensate the compass to lessen this error by the use of small magnets inside the compass. After this error has been minimized, a correction card is placed by the compass so the pilot can apply the appropriate correction as he flies.

The floating magnets align with the earth's magnetic field not only horizontally, but vertically as well. These fields are vertical at the poles where they enter the earth, and horizontal near the equator. The tendency of the compass card to align with this vertical component is called "dip," and must be compensated, so the card will not drag, or hang up. The card is slightly unbalanced to make this compensation. Most compasses used in the United States are compensated for about thirty degrees north latitude and are, for all practical purposes, satisfactory for most flight conditions.

This dip compensation keeps the compass from binding, but correcting one error gives two in its place - lead and lag, or northerly turning error, and acceleration error. The dip compensating weight is attracted by gravity to counteract the magnetic effect of the vertical component. In a coordinated turn, the gravity force acting on the weight is in line with the vertical axis of the airplane, but the magnetic force is still aligned with the center of the earth and is therefore exerting a pull on one side of the compass card. When banking from a northerly heading to turn either east or west, this side pull by the magnetic field causes the compass card to start to indicate in the opposite direction. Once the turn is established, the error will begin to decrease and will be zero at east or west. As the turn continues, the sideways pull acts in the same direction as the turn and the indication begins to lead the turn. When the airplane levels out, the compass is no longer affected by the side pull and indicates correctly.

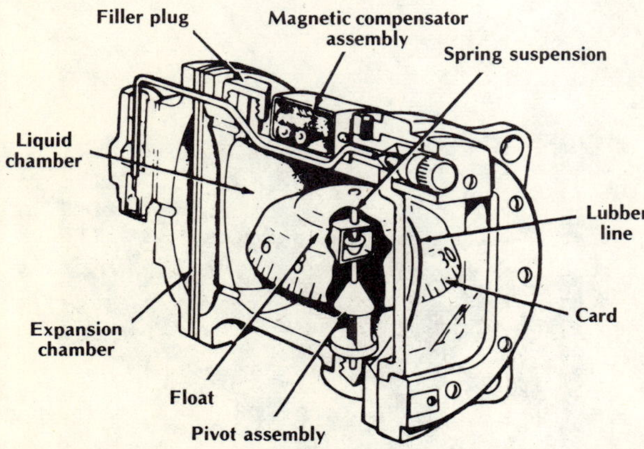

A magnetic compass consists, essentially, of a pair of permanent magnets soldered to a brass float suspended in a bowl of compass fluid.

- Figure 61 -

The weight also has an effect on easterly and westerly headings. When the airplane is flying on either of these headings and accelerates, by dropping the nose, the inertia of the dip correction weight causes the card to rotate in such a direction that it indicates a turn to the north. Deceleration causes an indication of a turn to a southerly heading.

g. Directional gyro:

The oscillations of the magnetic compass make it almost impossible to use for holding a heading in instrument flight, so the directional gyro has been developed. A small wheel having its weight concentrated around its rim is spun at high speed by either an electric motor or a blast of air. Unlike the rate gyro we have discussed, this gyro is suspended in a double gimbal, or a universal joint, so the airplane can turn completely around without causing any precession of the gyro. Attached to the gyro is a card or dial calibrated like that of a compass.

The directional gyro, while not a north-seeking instrument, may be set by reference to the magnetic compass so it can provide a stable and steady directional reference for the pilot.

- Figure 62 -

This dial indicates the position of the nose of the airplane relative to the spinning gyro.

The directional gyro must be set to agree with the magnetic compass at a time when its oscillations are minimum. Once set to the correct heading, it does not oscillate and provides the pilot with instantaneous and accurate information of any changes in his heading.

Special directional gyros slaved to the earth's magnetic field have become highly stable direction-indicating instruments, without oscillations or the lead and lag errors inherent in floating magnet compasses.

The flux valve generates a voltage whose phase changes as the nose of the airplane changes relative to the earth's magnetic field. This voltage causes a slaved directional gyro to indicate the airplane's magnetic heading.

- Figure 63 -

h. Gyro horizon:

The inner ear of man has a wonderful sensing system to keep him straight and level on the ground, but when the complex forces encountered in flight are involved, this system loses its effectiveness. Without some reference to the natural horizon, man is incapable of flying straight and level. To overcome this limitation, the artificial horizon has been developed which provides the pilot a horizontal reference and allows flight in conditions of no outside visibility. A small wheel similar to that used in the directional gyro spins in a double gimbal with its spin axis vertical. It senses rotation of the airplane about both the lateral (pitch) and the longitudinal (roll) axes.

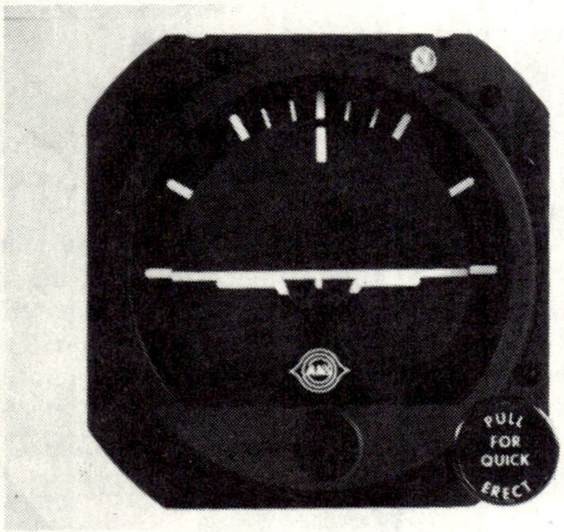

The gyro horizon provides a reference about both the pitch and roll axis of the airplane.

- Figure 64 -

The horizon bar remains level with the earth, and the miniature airplane, which is rigidly attached to the actual airplane, pitches and rolls about the bar. The little airplane is adjustable up and down to allow the pilot to get a nose level indication for any airspeed.

QUESTIONS:

62. What is the difference between a turn and slip indicator and a turn coordinator?
63. What causes deviation error in a compass?
64. What causes the lead and lag errors in a magnetic compass?
65. What does a directional gyro actually indicate?
66. About which two axes does a gyro horizon indicate rotation?

---

4. Avionics:

An entire book could and should be devoted specifically to helping the pilot get the most from his avionic installation; the purpose of this book, however, is merely to help understand the basic principles of these systems as they apply to the hardware, not to their operation. The treatment here simply introduces the equipment and its antenna.

a. Omni-range navigation equipment [VOR]:

The very high frequency omni-range equipment is a phase comparison type of navigation system, normally tied in with one of the communications transceivers.* After a navigation signal is received and amplified in the early stages of this equipment, it is sent into the omni circuits where it is separated into a variable signal and a reference signal which have been transmitted from the ground station. The omni bearing selecter (OBS) allows the pilot to shift the phase of the reference signal until it agrees with the variable signal. The OBS then indicates his magnetic direction from the station. Another phase shift circuit tells him whether, if he turns to the heading shown on the OBS, he will be going to and from the station. The signals for the VOR are horizontally polarized and are received on a horizontally mounted "V" dipole antenna, usually mounted on top of the cabin or the top of the vertical fin.

A horizontally mounted dipole antenna, above the cabin, receives the signal from both the VOR and the ILS localizer.

- Figure 65 -

The connection from this antenna into the instrument is made by coaxial cable, a special type of wire having a central conductor; an insulator; and outer conductor, which is a form of wire braid; and finally an outer insulator. When more than one VOR is coupled to a single antenna, a duplexer is installed between the instruments and the antenna. This matches the single antenna to the two receivers.

The communications equipment often incorporates the VOR, and the same antenna is used to receive voice communications from ground or other aircraft. Transmission is done on a separate antenna. This is a short whip or blade antenna, usually mounted on top of the cabin. There is normally one antenna for each transmitter.

Radio energy from the transmitter is radiated from a whip or blade antenna usually located on top of the cabin.

- Figure 66 -

b. Instrument Landing System:

[1] Localizer

The localizer uses the same receiver, antenna, and indicator as the VOR, but the principle of its operation is entirely different. The localizer transmits a pair of overlapping signals down the centerline of the instrument runway. Looking down the runway from the approach end, the signal sent out on the left side is modulated* at 90 cycles per second, and the signal on the right side is modulated at 150 cycles per second. These signals overlap by five degrees. The receiver circuit converts these two signals into DC voltages and compares them. The resulting signal indicates to the pilot which side of the centerline he is flying. When the localizer frequency is selected on the receiver, the localizer circuitry is automatically put into action.

[2] Glide slope:

A complete instrument landing system requires a glide slope. Its basic operational principle is the same as that of the localizer, in that 90 and 150 cycle modulation is used to establish the glide path. These signals are sent out with such a pattern that they overlap by about one degree, and this overlap forms about a two and a half to three degree approach angle with the ground. While the modulation is the same as that for the localizer, the carrier is in the ultra high frequency band and is received on a separate receiver and picked up on a different antenna. The UHF wave is much shorter than VHF, and the antenna is therefore much shorter - a short dipole, which may be mounted inside the airplane in the center of the windshield or on the top center of the fuselage.

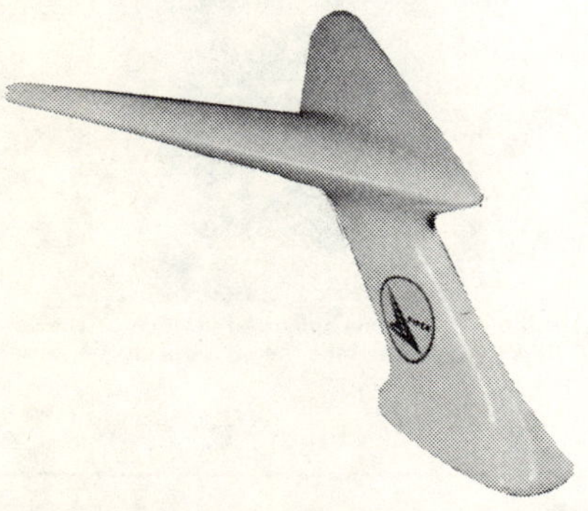

Piper

The Ultra High Frequency glide slope signal is received on a short dipole antenna mounted either on top of the airplane, in the nose, or inside the cabin.

- Figure 67 -

Since the glide slope uses a different transmitter on the ground and a different receiver in the airplane, it is important that the two be properly tuned so as not to receive the glide slope from one ILS and the localizer from another. To prevent this possibility, the two facilities are paired by the FAA, and when the localizer is tuned by the pilot, the glide slope paired to that frequency is automatically tuned. The glide slope antenna is connected to the receiver by coaxial cable, similar to that used by the VOR.

[3] Marker beacons:

The localizer nails down the centerline of the runway, the glide slope brings the pilot down at the proper approach angle, and the marker beacons locate him along the approach. These beacons all transmit on 75 megahertz with a highly directional antenna which may be received only directly above the antenna. The transmitter at the outer marker transmits a 400 cycle per second tone, and through a system of filters in the indicator causes a purple light on the instrument panel to illuminate in a series of dashes. When over the middle marker, a 1300 cycle per second modulation causes the amber light to light up with a series of alternate dots and dashes. Inner markers, or airways markers, are keyed with a 3,000 cycle per second signal which lights the white light.

Since the marker beacon receiver is a fixed frequency, low powered unit, it may be incorporated into other equipment such as the audio control panel. Normally only a sensitivity and perhaps a test switch are provided for this equipment. A small canoe shaped antenna mounted on the bottom of the airplane receives the marker beacon signal. It is connected to the receiver with the familiar coaxial cable.

The "canoe" antenna mounted on the airplane's belly is tuned to 75 megahertz to receive the signal from the marker beacon transmitter.

- Figure 68 -

c. Automatic Direction Finder:

VOR is the primary navigation facility used in the contiguous United States, but ADF is a highly useful back-up system. VOR operates in the VHF frequency range and is limited in its range to line of sight distances, but ADF operates in the low and medium frequency band and has a much greater range at low altitudes. VOR is concerned with the location of the airplane over the surface of the earth and is not affected by the heading of the airplane. ADF simply tells the pilot the bearing of the radio station from the nose of the airplane.

Two antenna are required for ADF operation, a loop and a long wire sense antenna. The loop may be one of two types, rotating or fixed. Rotating loops are often mounted under a plastic cover in the airplane tail cone or under a streamlined bump on the bottom of the fuselage. Fixed loops are much smaller and may be mounted flush with the bottom of the airplane or else extending only a half inch or so from the skin.

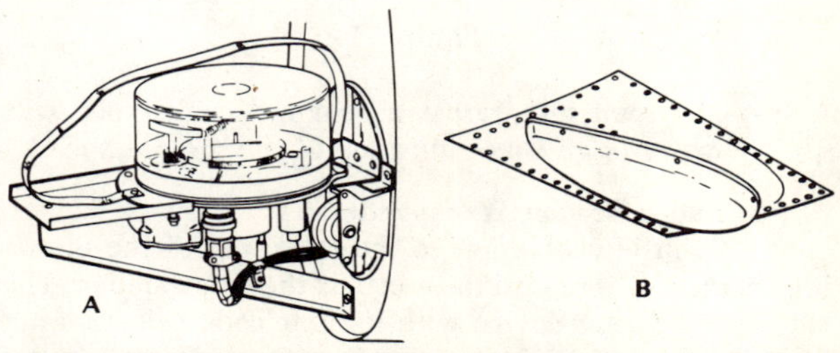

Automatic Direction Finders require a long wire sense antenna and a loop. The loop may be either a servo motor-driven, rotable loop [A] or a fixed, goniometer loop [B].

- Figure 69 -

With only a loop, the equipment could not tell whether the station is in front of the airplane or behind it, so a sense antenna is installed. This long wire, usually run from the top of the cabin to the vertical fin, is non-directional and adds its signal to that of the loop to provide directional sensing.

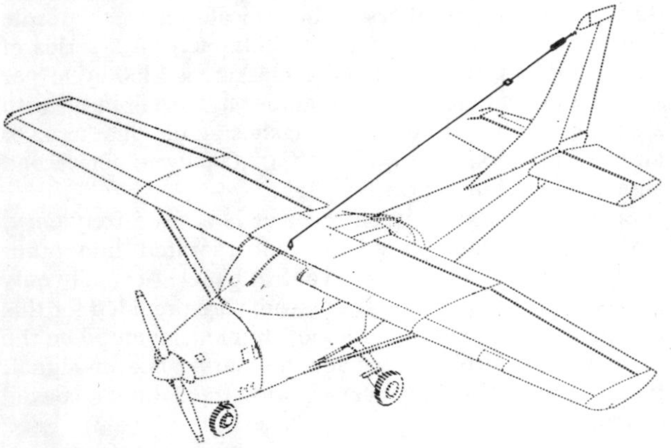

The long wire sense antenna used with the Automatic Direction Finder [ADF] removes the 180° ambiguity error and causes the equipment to indicate the direction of the signal relative to the nose of the airplane.

- Figure 70 -

d. Distance Measuring Equipment:

This installation is of a category known as pulse equipment. A signal is generated within the equipment and is transmitted to the ground from a small stub wire or UHF blade antenna beneath the airplane. When the signal is received on the ground, it is held for a few microseconds and re-transmitted on another frequency and received in the airplane by the antenna which transmitted it. The time for the total travel of the signal from its generation until it was received back into the equipment is translated into nautical miles to the station.

An Ultrahigh Frequency [UHF] blade antenna on the belly of the airplane both transmits and receives the signals used in the Distance Measuring Equipment [DME].

- Figure 71 -

A special circuit integrates the change in distance with time in order to give an indication of the ground speed.

e. Radar Beacon Transponder:

Of little visible use to the pilot, but of the utmost importance with regard to safety, is the transponder. This simple appearing device with its four code selectors and litle winking reply light transmits a response to the ground station identifying the radar return made by your airplane. This, like DME, is pulse equipment and connects to a small rod or blade antenna on the bottom of the fuselage. Altitude reporting is done by using a special altimeter which codes the transponder with the appropriate code for each altitude and replies with this code. Most transponders are capable of replying with 4096 different codes.

f. Emergency Locator Transmitter:

Back in the rear of the fuselage is a small transmitter we all hope never to use. It is the ELT. This small, self contained, pre-tuned transmitter is always armed, and an impact such as the airplane receives in a crash will trip a switch allowing the ELT to transmit a series of down sweeping tones over both the VHF and UHF emergency frequencies of 121.5 and 243.0 megahertz. ELT's have two items of concern for the pilot: First, be sure that they are **not** working. Tune the communications receiver to 121.5 mhz and listen. This should be done before starting the airplane and after shutting it down after flight. Second, the batteries must be replaced at a time interval equal to one half of their listed shelf life. The antenna for the ELT is usually a small flexible wire protruding from the rear of the fuselage or the tail, mounted in such a way that it is least likely to be damaged in a crash.

A thin, highly flexible antenna mounted in such a way as to most likely survive a crash is used to transmit the signal from the Emergency Locator Transmitter [ELT].

- Figure 72 -

QUESTIONS:

67. What kind of antenna is used with a VOR receiver?
68. What kind of antenna is used for the ILS localizer?
69. What kind of antenna picks up the ILS glide slope?
70. How does the pilot tune the ILS glide slope?
71. On what frequency are marker beacon signals transmitted?
72. What type of antenna is used with a marker beacon receiver?
73. What two types of antenna are required for an ADF system?
74. Name two types of ADF loop antenna.
75. What type of antenna is used for DME?
76. Where is the antenna for the radar beacon transponder mounted?
77. What type antenna is normally used with an emergency locator transmitter?

The cylinder with all its associated hardware is the power production section of a reciprocating engine. **Chrome Plate**

- Figure 73 -

SECTION IV: AN INTRODUCTION TO POWERPLANT SYSTEMS

A. OPERATING PRINCIPLES:

All licensed pilots are required to have a basic knowledge of aircraft engines, but in order to be more efficient in the operation of your engine, it is well to review what actually goes on inside the cowling.

For all practical purposes, engines found in modern light airplanes are of the four-stroke cycle, air cooled, horizontally opposed type. They are fueled with gasoline and ignited by two high tension magnetos. Lubrication is by mineral oil forced through a hollow crankshaft and distributed through drilled passages. The propeller is driven from the crankshaft either directly or through reduction gears which allows a higher engine rpm without excessive propeller tip speed.

Pistons fit into the cylinders and are moved in and out by connecting rods attached to the crankshaft. The cylinder heads contain two valves each, one between the combustion chamber and the intake system, and the other between the cylinder and the exhaust stacks.

To begin the cycle of operation, the piston is drawn inward with the intake valve open. An induction charge is drawn into the cylinder through the carburetor. Slightly after the piston reaches the bottom of its stroke and starts back out, the intake valve closes. This timing allows the valve to be open a maximum amount of time to draw in as much mixture as possible. Both valves are closed as the piston moves outward, compressing the fuel-air mixture. About twenty or thirty degrees of crankshaft rotation before the piston reaches the top of its stroke, two spark plugs ignite the mixture and the air begins to expand. Maximum pressure is obtained shortly after the piston reaches the top of its stroke and exerts a stout push on the piston, turning the crankshaft. Near the bottom of the power stroke, the exhaust valve opens and remains open all during the next outward stroke. The piston pushes all of the burned gases out of the cylinder and prepares for the next charge. The intake valve opens slightly before the piston reaches the top, so the low pressure caused by the rapidly departing and much cooled exhaust gases will help start the new charge into the cylinder.

B. MECHANICAL SYSTEMS:

1. Crankshaft:

The heart of an aircraft engine is its crankshaft, that special, high strength steel forging, which changes the in and out motion of the pistons into rotary motion and drives the propeller. The crankshaft is the heaviest single part of the engine. This fact has, to a great extent, dictated the configuration of aircraft engines. To get the shortest crankshaft possible, radial engines had all of their cylinders attached to a single throw of the crank. This provided a low engine weight for the horsepower produced, but as speeds became higher, the large frontal area proved too much penalty. In the evolution of engine configurations, the horizontally opposed engine has emerged as standard. In this arrangement, two cylinders attach to the crankshaft between each main bearing, and the shaft may be made relatively short for the power developed. Frontal area on these engines is small.

Crankshafts are usually hollow with oil under pressure

Crankshafts are made of the strongest alloy steels available and convert the in and out movement of the piston into rotary motion to drive the propeller.

- Figure 74 -

from the pump flowing through them. The main bearings support the shaft on a film of oil from the hollow shaft.

The propeller may be fastened directly to the crankshaft, or it may be geared and mounted on a separate propeller shaft. Geared engines usually have vibration problems because of the speed difference between the propeller and the crankshaft. For this reason, these crankshafts have torsional vibration dampers which rock back and forth as the shaft spins and absorbs the vibrations rather than putting stresses into the shaft.

2. Crankcase:

The cast aluminum alloy crankcase ties all of the engine together. The crankshaft is held centered in the case by babbitt lined bronze bearings; the camshaft is held in alignment with the crank by bearings cut into the case. The engine mounts into the airframe by attachments to the crankcase. The cylinders bolt to the case by studs and through-bolts which must provide enough strength to withstand the forces of the expanding gases in the cylinders. The front of the crankcase must not only be strong enough to withstand the thrust of the propeller, but also the tremendous side loads imposed by the shaft because of the gyroscopic forces of the propeller as the airplane pitches and yaws.

3. Cylinders:

The power developed in aircraft engines comes from the expanding gases pushing on the piston. The piston is actually a movable plug in the cylinder which allows the expanding gases to produce work. The cylinder of a modern aircraft engine consists of a high strength alloy steel barrel, having fins and a mounting flange machined on its outside. A cast aluminum alloy head is screwed and shrunk onto this steel barrel. The head has deep fins cast into it and ports for both the intake and exhaust valve. Two spark plugs screw into steel or bronze bushed holes in each cylinder head.

The cylinder walls are constantly rubbed by the piston rings and are naturally subject to considerable wear. To minimize this, some engines have the walls case-hardened with an ammonia treatment known as nitriding; others have the walls plated with porous chrome, whose surface contains millions of extremely tiny cracks. These cracks hold lubricant, and the chrome provides an exceptionally good wearing surface.

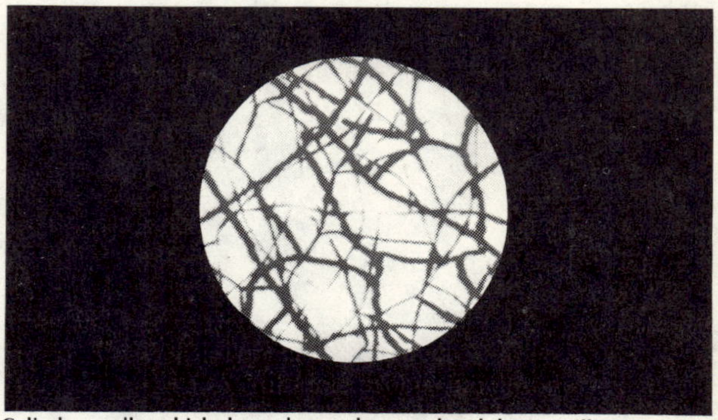

Cylinder walls which have been chrome-plated have millions of tiny cracks in their surface to hold lubricant.

Chrome Plate

- Figure 75 -

4. Pistons and rings:

Pistons for almost all modern engines are forged of high strength aluminum alloy. The need for strength is easily seen when we look at the movement of the piston. On take-off, for instance, if an engine with a five inch stroke is turning 2700 rpm, the piston travels 2250 feet each minute. In this minute, it has started from zero velocity, accelerated to its maximum rate and decelerated to a stop 5400 times. The piston is pushed down every other stroke by burning gases, and this heat and pressure causes the material and construction of the piston to be extremely critical.

There must be a tight seal between the piston and the cylinder wall. This is provided by cast iron piston rings. These rings are cut and expanded so they fit into the cylinder with a spring action which holds them tight against the wall and presents a good seal. Pistons usually have from three to five rings, the top rings specifically for sealing, and the lower rings to control the thickness of the oil film between the piston and the cylinder wall.

5. Valve Mechanism:

The fuel-air mixture must be introduced into the cylinder at the proper time and the burned gases exhausted on schedule. To do this, there is an intake and an exhaust valve, each actuated from the cam shaft through a push-rod and valve lifter. The valves are normally made of high strength nickel alloy steel and close against bronze or iron seats shrunk into the aluminum cylinder head. The valve stems ride in bronze valve guides. The exhaust valve must pass the exhaust gases which still have considerable heat energy in them, and on the larger engines, in order to dissipate the heat from the valve head, the valve stems are hollow and partially filled with metallic sodium. This melts and absorbs heat from the valve head, then sloshes up into the hollow stem and transfers it into the valve guide, then into the cylinder head. There are more fins on the exhaust side of the head than on the intake side, for the additional cooling required.

In order for the valve to operate efficiently, it must open at exactly the correct time, and it must lift off its seat the right distance. Any clearance between the valve stem and the cam shaft must remain exactly the same. If this clearance grows because of the cylinder expanding, the valve will open later than desired and will not open far enough. To prevent this, all modern aircraft engines use hydraulic valve lifters. This is a small hydraulic cylinder between the valve and the camshaft with a very accurately calibrated leak, so that as the valve is held open some fluid leaks out of the lifter.

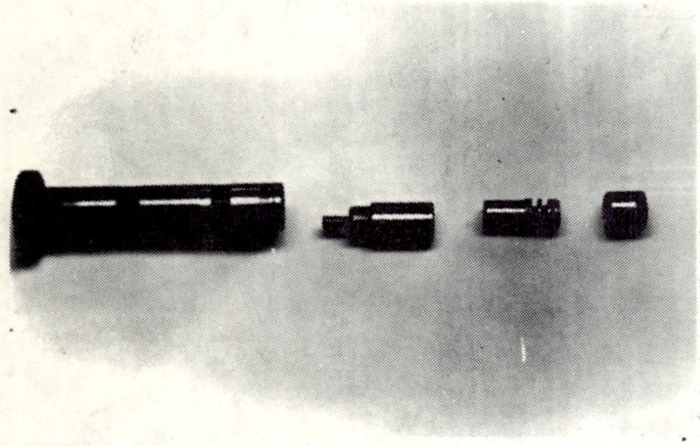

Hydraulic valve lifters maintain a condition of zero clearance between the rocker arm and valve stem. This minimizes wear and noise.

- Figure 76 -

When the valve is closed, the lifter refills with engine oil and is ready for the next lifting action. In this way, all of the clearance is taken up, regardless of the amount the engine dimensions change. There is much less noise and less wear because of this zero clearance.

6. Engine mounts:

The engine must be mounted into the fuselage or nacelle in such a way that it can vibrate without shaking the airframe. The mount must be strong enough to withstand all of the torque* loads imposed by the engine, and it must be able to hold the engine without sagging under the loads imposed when the airplane makes a hard landing.

Engines may be mounted with all of the mounts on the rear case, or with some of the mounts near the rear and others near the front of the engine. A popular engine mount is the dynafocal type. This is a rear mount in which all of the bolts point toward the center of gravity of the engine-propeller combination. In this way, all of the stresses are applied in the direction of the mounts' greatest strength.

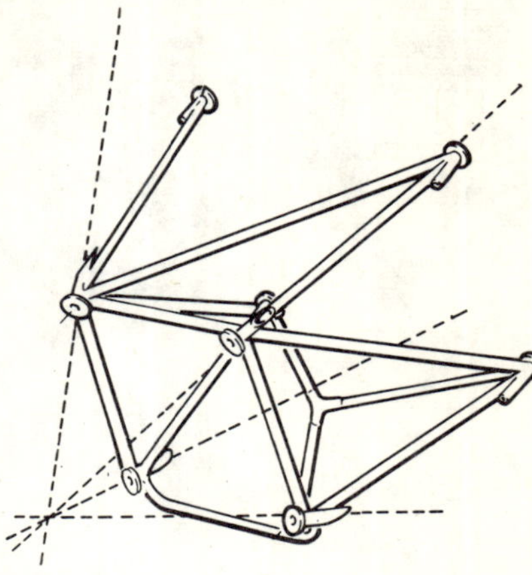

Dynafocal engine mounts have their mounting bolts pointed to the center of gravity of the engine and propeller combination. The rubber mounts operate under either tensile or compressive loads rather than shear or twisting.

- Figure 77 -

QUESTIONS:

78. Where is the piston relative to top center when normal ignition occurs?
79. Of what material are most aircraft engine crankcases made?
80. What is meant by a nitrided cylinder?
81. Of what material are piston rings usually made?
82. Why are some exhaust valves partially filled with metallic sodium?
83. Why do aircraft engines use hydraulic valve lifters?

C. COOLING SYSTEM:

An aircraft engine is a heat engine, meaning that the chemical energy in the fuel is converted into heat energy, and this into mechanical energy. A reciprocating engine is a very inefficient energy transformation device, transforming only about thirty percent of the power available in the fuel into actual horsepower at the crankshaft. The remainder of this released power must be dissipated as heat in such a way that it will cause no damage.

Most American-built automobiles use a liquid coolant to remove the heat from the engine, but aircraft engines cannot stand the weight penalty of this system, so they transfer the heat directly into the air. Early air cooled engines simply had the finned cylinders sticking out in the wind. This provided ample cooling for these relatively low powered engines, but modern engines, having so much more heat to dissipate, use pressure cooling. In this method, the cylinder fins are encased in sheet metal baffles which direct all of the air through the fins. The engine compartment is divided into two chambers, the upper has ram air in it, and the lower has a low pressure caused by a gill type opening at its rear. As the airplane flies, the high pressure air above the engine flows into the low pressure area, passing through the baffled fins where it picks up the maximum amount of heat from the cylinders. The pressure below the engine may be varied by the use of cowl flaps. The more the flaps are open, the lower the pressure and the more air there will be pulled through the fins.

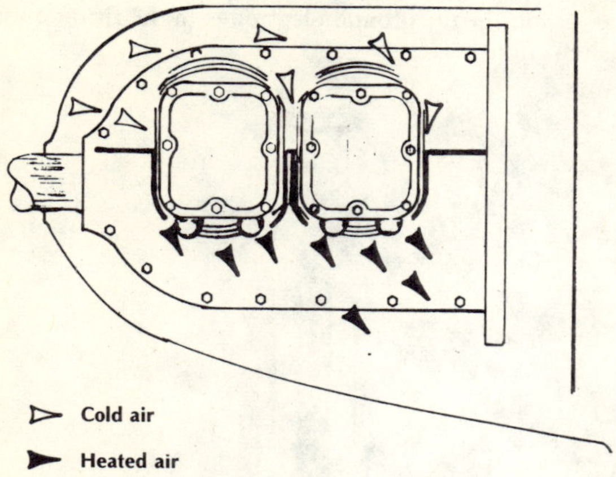

Modern aircraft engines are cooled by ram air which enters the upper part of the cowling, is forced between the fins by the baffles and out the bottom through the gill openings or cowl flaps.

- Figure 78 -

D. LUBRICATION SYSTEM:

The cooling system removes a large amount of waste heat from the outside of the engine, but there is still a considerable amount of heat inside which must be removed. This is one of the functions of the lubrication system. Another function of this system is the minimizing of friction which decreases the amount of heat generated.

1. Types of engine oil:

a. Mineral oil:

Straight mineral oil has been superseded, to a large degree, by other types for almost all operations except the first few hours of running in a new or a freshly overhauled engine. The other oils have so much better lubricating characteristics that there is not enough wear to properly seat the rings and bearings, so most manufacturers recommend straight mineral oil for the first twenty five hours of engine operation.

b. Detergent oil:

This oil has an additive which breaks the carbon bond to the metal and keeps the metal clean. Detergent oil is not used for aircraft engines now that better oils are available.

c. Ashless dispersant oil:

This is the most popular oil in use in piston driven aircraft today. Unlike detergent oils, it does not remove carbon already formed, but rather it prevents additional carbon sticking to the metal by keeping it in suspension.

2. Systems and components:

There are two types of lubrication systems found on modern engines: the wet sump, most commonly used now, in which the oil is carried in the lower part of the engine, in a sump, or oil pan; and the dry sump engine in which the oil is carried in an external tank. It is pumped through the engine and then returned to the tank. This type of system was extensively used with the radial engine, and to a lesser degree with opposed engines. Oil temperature is controlled by an oil cooler. If the oil is too hot, it flows through the core of the cooler; if it needs no cooling, it bypasses the cooler.

A fine mesh screen is installed at the outlet of the oil pump to collect any contamination in the system. If this screen becomes clogged, a by-pass valve is normally installed to pass the oil around the screen, providing some lubrication rather than allowing the engine to seize.

Full flow oil filters of pleated paper are installed in some engines. This filter may replace the internal screen and will trap contamination in its folds. Rather than being cleaned, this filter element is replaced at inspection time.

Oil filters are constructed of a special pleated paper. When replaced during an inspection, they are cut open to determine if any metal particles have been trapped.

-Figure 79 -

QUESTIONS:

84. What is the purpose of the baffles around the fins of an aircraft engine cylinder head?
85. How does an ashless dispersant engine oil keep the inside of the engine clean?

E. IGNITION SYSTEMS:

Perhaps the hardest working system on a modern aircraft powerplant is the ignition system. A hot spark must be provided at exactly the correct time to ignite the fuel-air mixture in the cylinders. The low density of the air at high altitude, moisture in the components, or oil or carbon on the spark plugs all collaborate to make the work of this system more difficult.

Since ignition is so vital, the FAA requires all certificated aircraft to have two independent ignition systems. This is not only for safety, but also for proper operation of the engine itself. Aircraft engines are required to release so much energy from the fuel that uniform burning of the mixture is important. To assure the best combustion, two spark plugs are installed in each cylinder, one on either side. The flame front therefore progresses across the piston head from both sides and meets in the middle. If one plug fails to fire, the flame must travel completely across the piston from one side. This takes longer and causes the mixture to get so hot that it may explode or detonate.

1. Magnetos:

Aircraft magnetos are AC generators with a current interrupter (the points) and a step-up transformer (the coil) to product its high voltage. This high voltage is carried into the engine cylinders where a spark is forced to jump across the gap in the spark plug. This spark, similar to a miniature lightning discharge, ignites the fuel air mixture so it can release its energy.

Aircraft magnetos use a permanent magnet geared to the engine, figure 80.

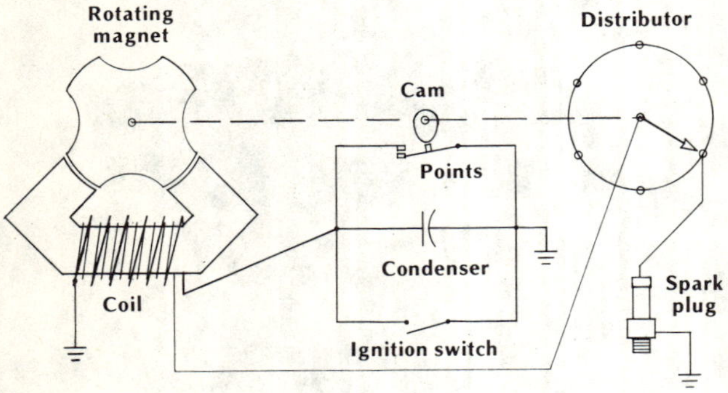

The magneto is an AC generator with a set of current interrupters in the primary circuit. When the points are opened by the engine-driven cam, the primary current stops. This causes the magnetic field to collapse and induce a high voltage in the secondary winding.

- Figure 80 -

The lines of magnetic flux from this magnet cut across the primary winding in the step-up transformer. When the greatest amount of current is flowing in this winding, the breaker points open and the magnetic field around the primary coil collapses instantly. This rapid collapse induces a high voltage in the secondary winding, and current flows through the distributor, to the proper spark plug, and jumps the gap to ground, thus igniting the mixture.

A condenser in parallel with the points absorbs some energy from the magnetic field as the points open, and aids the rapid collapse. Current flows through the points as they begin to open, and were it not for the condenser, an arc would form across the points, severely burning or pitting them. The condenser absorbs this flow and protects the points.

The operation of an ignition switch for a magneto is opposite to that of an ordinary electrical switch. When the magneto is "hot," or operating, the switch is open so all of the primary current must flow through the breaker points. When the switch is "Off," it is closed, and the primary current goes directly to ground, bypassing the points.

2. Spark plugs:

So little and so unobtrusive is the spark plug, yet so important. Almost all spark plugs used in modern aircraft engines are shielded. The entire ignition system is encased in a metal shield which collects all of the radiated electrical energy and passes it to ground, rather than allowing it to cause radio interference.

A ceramic sleeve fits into the metal case insulating the center electrode from the shell. A special corrosion resistant gasket seals the center electrode against leakage of the high-pressure, high temperature gases in the cylinder during the process of combustion.

There are two types of plugs in use. One, the massive electrode plug with a large diameter copper-cored electrode in the center of the ceramic insulator, and two, three, or four heavy ground electrodes in the firing end of the shell, figure 81.

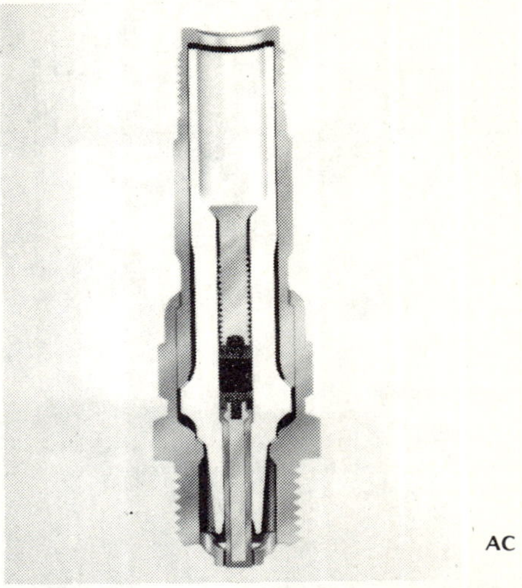

Massive electrode spark plugs have a copper-cored, nickel alloy center electrode and two or more massive ground electrodes.

- Figure 81 -

As the power output of piston engines has increased, more lead has been used in the fuel to prevent detonation. This lead is not all scavenged from the cylinders, and some forms deposits in the firing cavity of the plug. To minimize the effect of these deposits, spark plug manufacturers have developed plugs with fine wire electrodes. These plugs have small diameter platinum or iridium electrodes, figure 82.

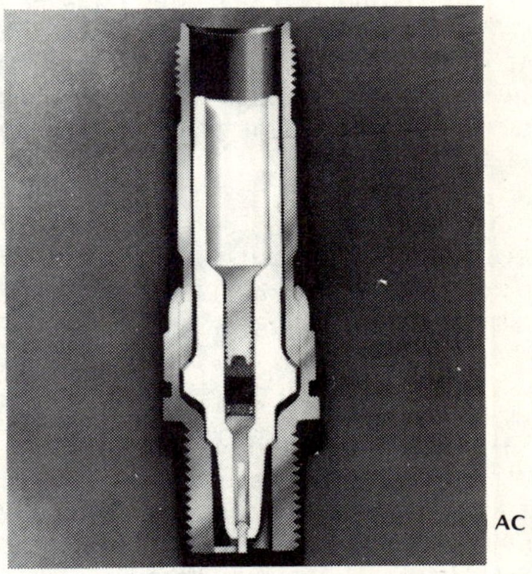

An "all weather" spark plug with fine wire electrodes.

- Figure 82 -

Fine wire plugs may be operated at higher temperatures than massive electrode plugs, and thus burn off the lead deposits.

The shielding on the spark plug leads acts as a capacitor and stores electrical energy. When the plug fires, the stored energy is released and causes a large flow of current which erodes the electrodes. To minimize this erosion, resistors are built into some plugs to limit the peak current.

There are many different kinds of spark plugs on the market today. Most of them will fit into the hole in the cylinder, but they have different operating characteristics. The FAA lists in the engine Type Certificate Data Sheets all of the spark plugs approved for a given engine. The use of a plug not having FAA approval is quite likely to cause engine malfunction.

3. Ignition harness:

The wiring between the magneto and the spark plug is one of the most important parts of the ignition system, yet it is one of the least considered components. Each individual lead attaches into the magneto distributor block and to the spark plug with threaded terminals. The wire is usually stranded stainless steel, covered with a silicone rubber insulator. Around this insulator is a highly conductive metal shielding, grounded at both ends. Any radiation of electrical energy is intercepted by this shielding and conducted to ground. Almost all modern ignition harnesses are covered with teflon or other tough, abrasion resistant materials.

The spark plug end of the harness is the most critical. It terminates with a spring which holds pressure against the center electrode and is held centered in this barrel by a phenolic, ceramic, or silicone rubber "cigarette."

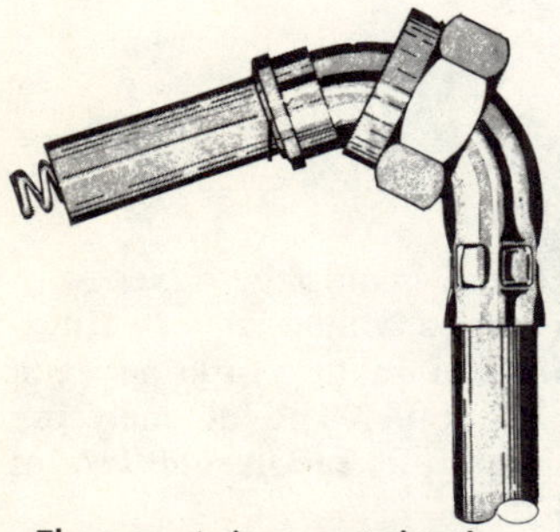

The magneto is connected to the center electrode of the spark plug by a spring-loaded connector, held in the plug by a ceramic, phenolic, or silicone rubber "cigarette."

- Figure 83 -

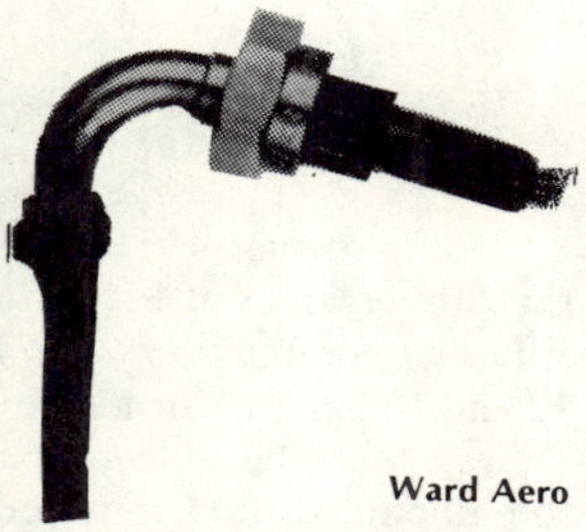

Ward Aero

"All weather" spark plugs have their leads sealed into a recess by a resilient seal on the harness lead.

- Figure 84 -

Any moisture in the harness will form a short circuit, and the engine will miss and lose power. To allow the ignition system to operate in heavy rain conditions, the "all-weather" spark plug and harness has been designed. This plug can be recognized by its large barrel and recessed ceramic insulator. A seal like a large rubber stopper excludes all moisture from the plug and harness.

4. Aids to starting:

a. Impulse coupling:

Automobiles use battery ignition systems in which the intensity of the spark is not dependent upon the speed of rotation of the engine, and an adjustable spark system delays the spark for starting until the piston is on its downward stroke. Aircraft engines use magneto ignition, in which the speed of rotation determines the intensity of the spark. To provide a spark that is both hot and late for starting, an impulse coupling is used on one or both magnetos.

This coupling uses a heavy spring similar to an overgrown clock mainspring. When the engine is turned over by the starter, arms on the coupling engage stops on the magneto housing and hold the rotating magnet until the engine has rotated well past the normal firing position. By the time the piston passes over top center and starts down, the spring is wound up tight, and the arms slip off the stops. The rotating magnet is spun by the spring and produces the required hot, late spark. As soon as the engine starts, centrifugal force acts on counterweight arms and pulls them in so they no longer contact the stops.

b. Induction vibrator:

Disadvantages of the mechanical impulse coupling have brought about the development of a highly effective induction vibrator system to provide the starting spark. The left magneto of an engine equipped with this system has two sets of breaker points, one on either side of the cam. These points are arranged in such a way that one set, the normal, or run points, opens when the piston is about thirty degrees before top center. The retard points open about the time the piston passes over top center.

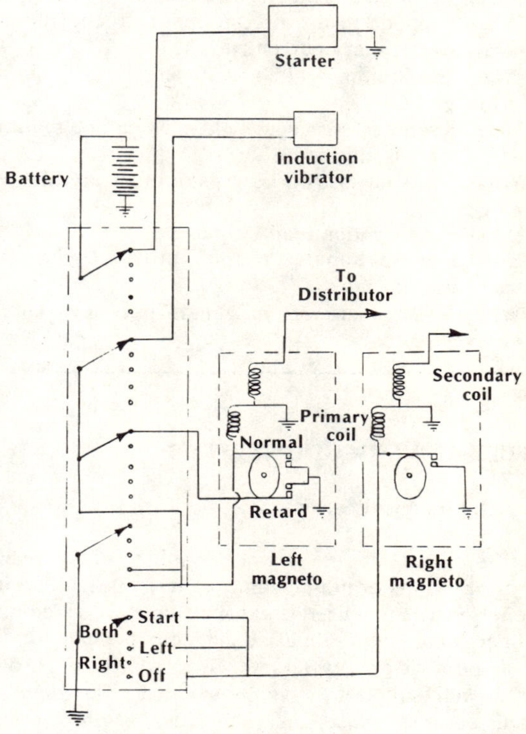

The hot, late spark for starting comes from an induction vibrator sending pulsating direct current through a set of points which open later than the normal, or "run", points.

- Figure 85 -

As shown in figure 85, when the ignition switch is placed in the start position, the following events occur:

(1) The right magneto is grounded so it will not interfere with the starting procedure.

(2) The battery is connected to the starter solenoid, allowing current to flow in the starter and crank the engine.

(3) The induction vibrator is energized and starts chopping up the direct current, sending pulsating DC into the left magneto. Since both sets of points are closed, all of this current goes to the ground.

(a) The normal or run points open first, but since the

retard points are still closed, the current goes to ground through them.

(b) When the retard points open, there is no route for the current to ground except through the primary coil of the left magneto. This pulsating DC flowing through the coil induces a high voltage in the secondary which is connected to the spark plug through the distributor. The spark plug gets a good hot, late spark. When the engine starts and the switch is placed in the "Both" position, all of the contacts are open and no current flows to either the starter solenoid or the vibrator. The retard points are out of the circuit, and both left and right magnetos operate normally. When the switch is turned "Off," the two lower contacts close, grounding both magnetos.

QUESTIONS:

86. For what reason, other than safety, do aircraft engines use dual ignition?
87. What happens in an aircraft magneto the moment maximum primary current flows?
88. When the ignition switch is closed, is the magneto "On" or "Off"?
89. Why do some aircraft spark plugs use a fine platinum wire for their electrodes?
90. What is the purpose of the resistor in an aircraft spark plug?
91. What is an ignition lead "cigarette"?
92. How does an impulse coupler provide a hot, late spark for starting?
93. Why do some aircraft magnetos have two sets of breaker points?

F. FUEL AND INDUCTION SYSTEM:

1. Aviation gasoline:

The aircraft engine is a heat engine, which derives its power by transforming chemical energy in the fuel into heat energy. The fuel used is gasoline which has a nominal heat energy content of about 20,000 Btus per pound. This large amount of energy and the availability of gasoline make it well suited for use in aviation. The demands modern engines place on their fuel require certain additives to be used which allow higher cylinder pressures before ignition occurs. Tetraethyl lead retards the burning rate and allows higher pressures without detonation, but this lead has a tendency to foul spark plugs and cause valve stem erosion, so a second additive, ethylene dibromide, is used to help scavenge the lead.

Aircraft gasolines are rated according to their anti-detonation (anti-knock) characteristics, and the proper fuel must be used in an engine. This classification is known as its octane rating. Aviation gasoline uses a dual number system, such as 80/87 and 100/130. This simply means that, with a lean, cruise mixture, the fuel behaves as one having the lower rating; with a rich mixture such as used for take-off, its anti-detonation characteristics are the same as a fuel having the higher rating.

In the past, there have been several grades of fuel used, but the economic trend is to reduce this as much as possible. Today, three grades of gasoline are produced, and efforts are being made to reduce this to one. 115/145 is used primarily for engines with very high compression ratios, mainly by the military. We will not discuss it more than to say it is dyed purple for identification.

80/87 is used for the lower compression engines in general aviation. It is allowed to contain a maximum of 0.5 milliliter of tetraethyl lead per gallon of fuel and is dyed red.

100/130 is the most popular fuel in use today, having been used since before World War II for the higher powered engines. 100/130 is dyed green and may contain as much as 4.6 milliliters of lead per gallon. Stocking both 80/87 and 100/130 av. gas. is uneconomical, and the petroleum industry is trying to satisfy both requirements by the use of a low lead content 100 octane fuel. The lead content of this fuel is two milliliters per gallon.

Fuel having an octane rating lower than required by your engine will cause detonation and should **never** be used. Prolonged use of fuel with a higher lead content than required may cause valve and spark plug problems, so consult your A&P before using any fuel other than that specified in your owner's manual.

Fuel additives, other than those put in by the refiner, are to be avoided. The engine manufacturer has made in-depth study of fuels and their additives and naturally recommends that which will give the best results in his engine.

2. Fuel Metering:

Liquid fuel will not burn, so it must be mixed with the proper amount of air to form a combustible mixture. Efficient combustion occurs with a mixture ratio between 12 and 16 parts of air to one part of fuel. The fuel metering system must be able to maintain this ratio regardless of the amount of fuel required.

There are three types of fuel metering systems in use for aircraft engines: the float carburetor is still used on the lower powered engines, as it has been since World War I; the pressure carburetor which came into its own during World War II; and the fuel injection system, which is perhaps the most popular fuel metering system in use today. All three types of metering devices will be discussed.

a. Float carburetors:

Float carburetors meter the fuel into the air by a pressure drop caused by the air the engine breathes. All of this air must pass through the venturi.

- Figure 86 -

This simple device has five systems which operate to perform the metering function for the smaller engines.

[1] Main metering system:

Fuel from the tank enters the float bowl of the carburetor of Figure 86 where a constant level is maintained by the float controlled needle valve. All of the air that enters the engine must flow through the venturi.* The amount of this air is controlled by the throttle butterfly valve. The reduction in area caused by the venturi requires the air to speed up, and as the airspeed increases, according to Bernoulli's principle, the pressure at the discharge nozzle drops. Since the pressure at the nozzle is lower than that in the float bowl, fuel will be forced from the bowl, through the main metering jet, and out the discharge nozzle. At relatively low airflows, only a small amount of fuel is pulled out the discharge, but as the airflow increases, the amount of fuel discharged increases more than the air and the mixture ratio becomes richer. To prevent this, an airbleed is introduced which draws air from outside the venturi and mixes it with the fuel just below the discharge nozzle. This airbleed serves two functions: First, it maintains a relatively constant fuel-air mixture ratio as the airflow increases, and, secondly, it aids in breaking up the fuel for better vaporization.

[2] Idle system:

When the airflow is insufficient to cause fuel to flow out of the main discharge nozzle, a separate system must be used. This is almost like another carburetor having its discharge at the edge of the throttle valve. This is the point of lowest pressure at idle, and the fuel is drawn from the main system up through the idle metering jet to this discharge. An airbleed in the idle system aids in atomizing this idle fuel.

[3] Acceleration system:

If the throttle is opened suddenly from idle, there is likely to be a momentary lag between the time the engine loses the idle system fuel, and the time the main metering system begins to function. To minimize this lag, a pump is attached to the throttle to inject fuel into the air flowing into the engine, providing a momentarily rich mixture which prevents any hesitation.

[4] Mixture control:

A manually controlled valve in the passage between the float bowl and the discharge nozzle limits the amount of fuel allowed to flow. Since the engine is concerned with the ratio of the fuel to the air by **weight** and the carburetor meters the fuel according to the **volume** of air flowing into the engine, provision is made to decrease the amount of fuel as the airplane ascends into the less dense air.

[5] Power enrichment system:

At full power an aircraft engine releases so much heat energy from the fuel that damage is likely to result. Part of this heat may be removed from the cylinder heads by enriching the mixture so there will be an excess of fuel. Heat is absorbed in changing this extra liquid fuel into a vapor - the heat which would damage the engine.

There are several ways of adding this fuel into the airstream, but one of the more simple systems simply uses a valve and a jet in parallel with the main jet of the carburetor. When the throttle is fully opened, this valve opens and additional fuel is metered into the air. The popular Marvel-Schebler carburetor uses a restrictor actuated by the throttle in the main airbleed. At all conditions below full throttle, the airbleed is open, admitting air into the fuel being discharged. When the throttle is opened wide, this airbleed is restricted, causing more fuel to be pulled out of the discharge.

b. Pressure carburetors:

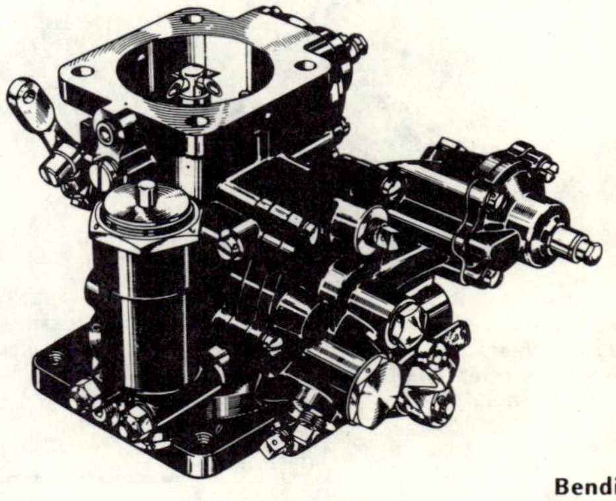

Bendix

Pressure carburetors meter the fuel into the air by a pressure differential caused by both the volume and density of the air used by the engine.

- Figure 87 -

When more precise metering is required than can be provided with a float carburetor, a pressure carburetor may be used. An air pressure differential is used to control the amount of fuel metered. This differential is between ram air pressure, which senses air density, and venturi pressure, which senses the velocity. These pressures determine the mass of air entering the engine, and they control a poppet valve which meters the fuel proportional to the weight of air taken into the engine.

A needle valve between the two air chambers allows the pilot to decrease the pressure differential to lean the mixture as the air becomes less dense at altitude. An automatic mixure control is placed in parallel with the manual control on some models, to keep the mixture constant as altitude changes.

These carburetors have enrichment systems to add more fuel at full throttle for cooling, and an accelerator system to momentarily enrich the mixture when transitioning from idle to the main metering system.

c. Fuel injection systems:

When we hear of a fuel injection system, we immediately think of a system such as used in a Diesel where fuel is injected directly into the combustion chamber under high pressure. The constant flow injection system used with modern aircraft is not such a system. It is very similar to a pressure carburetor. One of the main reasons for using an injection system is its evenness of fuel distribution. A carbureted engine has its fuel mixed with the air at the beginning of the induction system, and because of the different lengths of induction tubes and the pulsating flow of the mixture, all cylinders do not receive the same amount of fuel. An injection system delivers the fuel from individual nozzles just outside the intake valves and all cylinders receive the same amount of fuel.

There are two systems in popular use today, the Bendix RSA and the Teledyne Continental systems.

[1] Bendix RSA system:

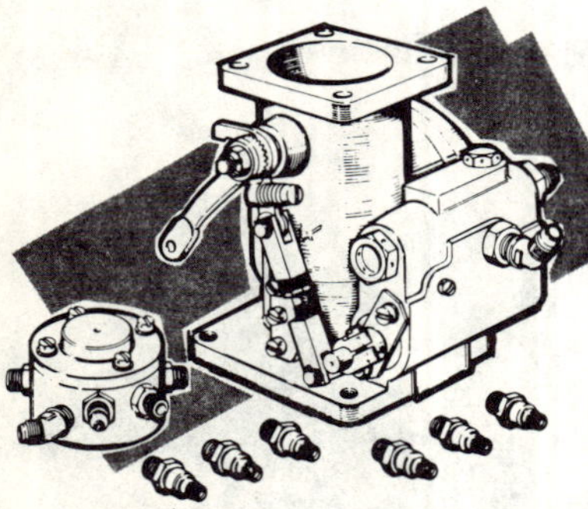

Bendix

The Bendix fuel injection system meters fuel in much the same way as a pressure carburetor. The principle difference between the two is the point of discharge. Injectors discharge their fuel near the engine intake valve through injector nozzles.

- Figure 88 -

The air metering force is measured in the same way as with the pressure carburetor, and this force controls a valve in the fuel flow line to the engine. One principle difference between this system and the carburetor is the type of fuel control unit used. A flat plate varies size of the fuel passage in the mixture control. It is closed at "Off" and wide open for "Full Rich". When wide open, metering is done by the main jet. After leaving the main jet, fuel flows through the throttle fuel valve which may also limit the flow of fuel to the engine. The throttle fuel valve moves with the air valve so they do not affect the fuel-air mixture ratio. After leaving the fuel control unit, the fuel flows to a flow divider which distributes it uniformly to all cylinders. Air is drawn into the liquid fuel in the injection nozzles through airbleed holes to aid in vaporization.

[2] Teledyne-Continental system:

The end result of the Teledyne-Continental system is the same as that of the Bendix, but the methods of achieving it are different. The Continental system uses no air metering forces; the fuel pump is the heart of this system. Figure 89 shows how the fuel is taken into the pump and moved by the eccentric vanes. The orifice allows some fuel to return to the inlet side of the pump, and this return causes the pump outlet to vary directly with the pump speed, and thus the engine speed. For idle speeds, where the orifice alone would not provide enough pressure for engine operation, the relief valve restricts the

Legend

1 - Fuel tank
2 - Boost pump
3 - Main strainer
4 - Swirl chamber
5 - Vane type pump
6 - Orifice
7 - Relief valve
8 - Pump bypass valve
9 - Strainer
10 - Mixture control
11 - Metering jet
12 - Throttle fuel valve
13 - Throttle air valve
14 - Manifold valve
15 - Nozzles
16 - Flow meter

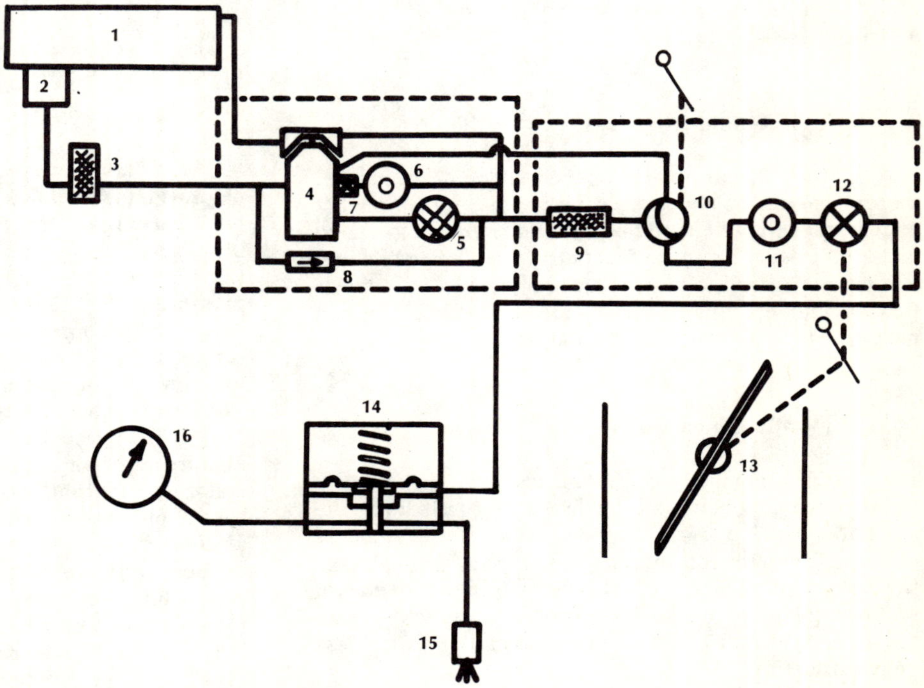

The Teledyne-Continental fuel injection system meters its fuel as a function of engine RPM by a special pump. There is no venturi in this throttle body.

- Figure 89 -

by-passed fuel and allows the pump to provide the proper amount of fuel for idling. Two of the more important adjustments of this type of system are the high and low unmetered fuel pressure. These adjustments are made by varying the size of the orifice and the pressure on the relief valve spring.

Another difference between the two systems is the mixture control. The Bendix system shuts off all flow of fuel from the fuel control for idle cut-off. The Continental mixture control is a variable selector valve. For cut-off the fuel is by-passed back to the inlet side of the pump. For full rich, all of the fuel is directed to the engine. Leaning is accomplished by varying the amount of fuel returned.

QUESTIONS:

94. Why is tetraethyl lead added to aviation gasoline?
95. What is meant by low lead 100 octane aviation gasoline?
96. What causes the mixture ratio of an aircraft engine to grow richer as the airplane goes to altitude?
97. Why do aircraft engines operate on a richer than normal mixture for full power?
98. What is one real advantage of a fuel injection system over a carburetor?
99. What forces does a Bendix fuel injection system use to determine the amount of fuel injected into the engine?
100. What does a Continental fuel injection system use to determine the amount of fuel injected into the engine?

---

d. Induction system:

Some parts of an engine are vital, yet are taken too much for granted; the induction system is such a part. The carburetor or fuel injection system mounts on the bottom of the engine and is connected to the cylinders through thin wall pipes. In some engines, these pipes are routed through the oil sump to warm the mixture and aid in the vaporization of the fuel. The design of these systems is quite complex because of the importance of keeping the fuel-air mixture constant in all cylinders. The induction system may be thought of as similar to a trombone. The length of the air passages has a marked effect on the pressure of the air pulses as they leave the bell. In an aircraft engine, for uniform fuel-air mixture, the pipes should all be exactly the same length, and since this is almost impossible to achieve, a compromise has been made to gain the best performance within the limits of the design. Maintenance-wise, this system must be kept absolutely tight. The slightest air leak into the system will cause a cylinder to operate overly lean and will in all probability cause detonation and severe engine damage.

e. Turbochargers:

The utility of an airplane is increased when it is enabled to fly above the weather. In order to do this, the plane must have additional air pumped into the engine. As we saw earlier, about seventy percent of the heat energy of the fuel is lost, a great deal of it in the exhaust. In order to recover some of this energy, a turbine may be fitted into the exhaust to drive an air compressor. Compressed air from this pump is directed into the carburetor where its increased density allows the engine to develop its rated sea level power up to a considerable altitude. If this air pump were to increase the intake manifold pressure too much, detonation and engine failure would result; so a turbocharger controller must be used to control and limit the pressure. This control senses the manifold pressure and operates a hydraulic actuator to direct the exhaust

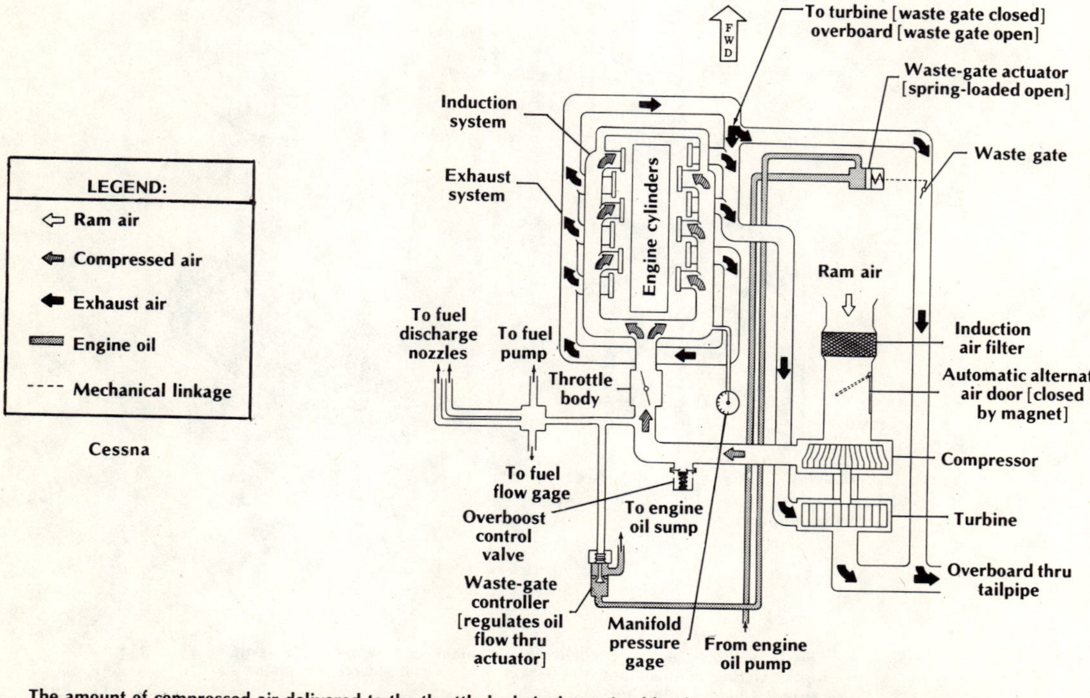

The amount of compressed air delivered to the throttle body is determined by the manifold pressure and is controlled by a hydraulically actuated waste gate. The more air needed, the faster the turbocharger turns.

- Figure 90 -

gases either overboard or through the turbine, When the pressure rises too high, more exhaust gas is dumped overboard and the turbine slows somewhat.

QUESTIONS:

101. What is an advantage of mounting the carburetor on the bottom of the oil sump?
102. What drives a turbocharger compressor?

G. PROPELLERS:

The propeller converts the torque of the engine into thrust to move air. As the air moves back, the airplane moves forward. This oversimplification belittles the importance of the propeller, however. Looking at the propeller carefully, we see that it is a rotating airfoil whose cross section changes from the hub to the tip. This airfoil must absorb all of the power of the engine, allowing it to develop the proper RPM.

Early propellers were carved from laminated birch blanks, but since World War II, almost all fixed pitch propellers have been made of forged aluminum alloy. Metal has the advantage of not being affected by the atmospheric conditions, and repairs can be made in the case of minor damage.

Modern high performance aircraft can no more tolerate the inefficiency of a fixed pitch propeller than an automobile could operate with only one gear. In the quest for more efficient propellers, the two position propeller was developed, allowing the pilot to use the low pitch for take-off. The blade bites the air with a lower angle and allows the engine to develop its full RPM and therefore its full power. This is the same as an automobile starting off in low gear. When a power reduction is made, the propeller is put into higher pitch so it can pull the airplane farther through the air for each revolution, and the engine makes fewer revolutions per minute.

Replacing the two-way control valve in the two position propeller with a centrifugal governor produced a constant speed propeller. For take-off, the propeller is set for full low pitch so the engine can develop full power and maximum RPM. When a power reduction is made, the propeller control is pulled back to a specific RPM. The governor holds the engine at this speed. If the nose drops, decreasing the load on the propeller, rather than the engine speeding up, the governor increases the pitch of the propeller and maintains the same RPM. If the nose rises, increasing the propeller load, the pitch will decrease.

There have been several methods used to change the pitch on constant speed propellers: electric motors, hydraulic actuators, compressed air, even hand cranks. The one method which has survived for use by the smaller general aviation aircraft is the hydraulic control, using engine oil as the fluid. Counterweights are used on some propellers to produce a force in the control action.

As the propeller spins, centrifugal force acting on the blades tends to cause them to streamline into the plane of rotation. This force drives the blade toward a lower pitch. Some propellers counteract this centrifugal twisting moment (CTM) with a counterweight. Centrifugal force acting on these weights tends to rotate the blade toward high pitch, Figure 91.

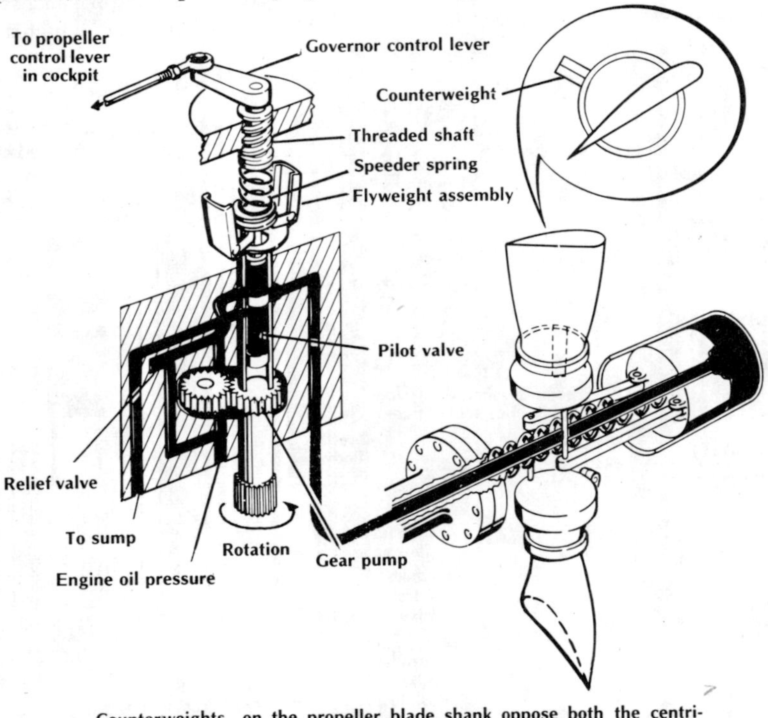

Counterweights on the propeller blade shank oppose both the centrifugal twisting moment and the action of governor oil on the piston to cause the blades to go into high pitch.

- Figure 91 -

McCauley

Propellers not using counterweights normally use a balance of forces between CTM and a spring, tending to move the blades into low pitch, balanced by oil pressure on a piston moving the blades into high pitch. To increase the pitch, the governor directs oil into the propeller and overcomes the spring force and the CTM, moving the blades into a higher pitch. To increase the RPM, the governor bleeds oil from the propeller, and the CTM and spring force move the blades to a lower pitch.

When counterweights are used, the action of the engine oil is reversed. CTM still tends to place the blades in low pitch, but this is overcome by a stronger centrifugal force on the counterweights, so the net force tends to move the blades toward high pitch. When the governor directs engine oil into the propeller, the blades move toward low pitch; and when oil is bled out, forces on the counterweights move the blades into a higher pitch.

The propeller governor uses as its speed sensor a pair of fly weights. These spinning weights tend to lift a hydraulic pilot valve. The action of the flyweights is opposed by a speeder spring. The pilot controls the tension of this spring with the propeller pitch control. For take-off, the control is full forward and tension on the spring is greatest. The engine therefore must develop its maximum RPM before the valve rises. At power reduction, the tension of this spring is reduced, and the engine speed stabilizes at a lower RPM.

QUESTIONS:

103. What does a propeller governor vary in order to maintain a constant RPM?
104. Do the forces on the counterweights of an aircraft propeller tend to move the blades toward high or low pitch?

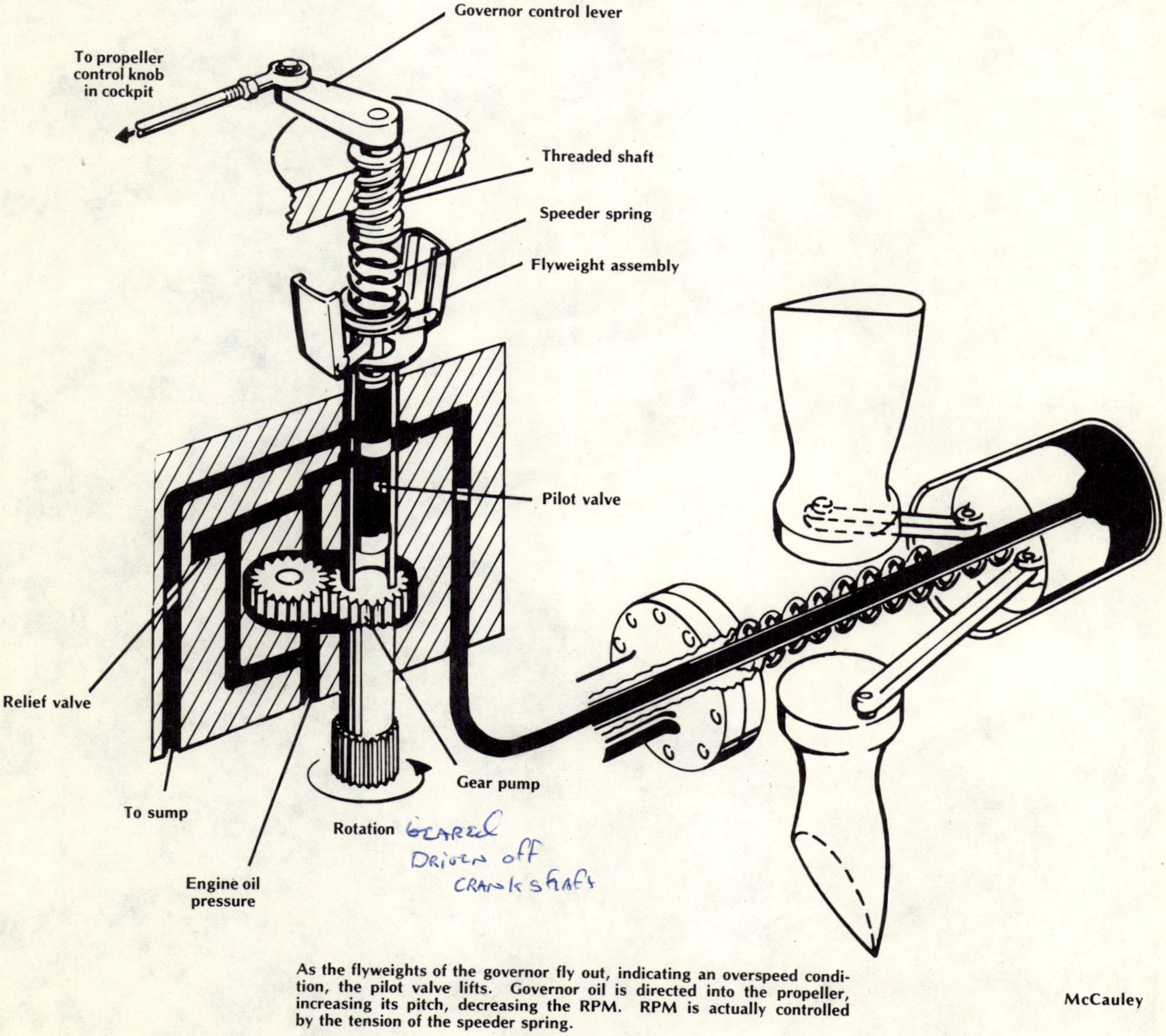

As the flyweights of the governor fly out, indicating an overspeed condition, the pilot valve lifts. Governor oil is directed into the propeller, increasing its pitch, decreasing the RPM. RPM is actually controlled by the tension of the speeder spring.

McCauley

- Figure 92 -

SECTION V: WHAT THE I.A. LOOKS FOR ON AN ANNUAL INSPECTION

Aircraft maintenance may be broken down into three categories: inspections, repairs, and alterations. Most of the work done on modern aircraft is inspections. You, the pilot, are required to conduct an adequate pre-flight inspection before each flight. The manufacturer normally recommends a twenty-five hour inspection, and the FAA requires an annual inspection for each certificated aircraft, or, if the aircraft is operated for hire, an inspection of the same depth every 100 hours of flight.

Lest there be any misunderstanding regarding inspections, the FAA makes no provisions for any person not holding a mechanic's license to inspect an airplane under the supervision of a licensed mechanic. A licensed mechanic may personally supervise an unlicensed person doing repair work, or making an alteration, but the responsibility for this work and the inspection for its return to service cannot be delegated by the mechanic. A pilot or other person not holding a mechanic's license may assist the mechanic on an inspection by removing and replacing inspection covers, and doing certain of the time consuming functions, but **the inspection itself must be conducted by a person holding the appropriate licenses.**

Regardless of how familiar a technician may be with a particular airplane, he is required to use a check list for the inspection. This list may be prepared by the airplane manufacturer; or it may be one compiled by the inspector. Almost all manufacturers furnish adequate inspection guides, and the inspector will normally add items peculiar to the specific installation, or may make modifications appropriate to the local operating conditions.

In this section of the book, we will consider the details of an annual inspection as we watch the inspection of a typical light single engine retractable landing gear airplane.

A. PRELIMINARY PAPERWORK:

Some of the more important functions of an inspection actually occur before the inspector touches your aircraft. When the airplane is brought in the shop, the work order is made out and signed by you, the owner, stating exactly what work is to be done to the airplane. A cost estimate is usually given which includes the cost of the inspection and an hourly charge for labor required to work off any discrepancies found in the inspection.

The inspector checks the Airworthiness Directives to see if there are any outstanding against the airframe, engine, or any component. He checks the Inspection Aids to find what troubles have been found in other aircraft of this same type, and he checks the manufacturer's service bulletins for changes recommended by the manufacturer for better service of your airplane. The aircraft records are then brought in and examined for total time on the airframe and engine, and the record of A.D. notes complied with. The equipment list is checked against the actual installed equipment and the weight and balance information checked to see that all of the proper computations are in the records. Any repairs or alterations which have been made to the airplane should be recorded on an FAA Form 337 and kept with the airplane. The aircraft and engine logbooks should record all inspections, repairs, or alterations which have been made.

In the airplane must be the Aircraft Airworthiness Certificate (FAA Form 8100-2), the Aircraft Registration

Certificate (FAA Form 8050-3) and the Aircraft Radio Station License (FCC Form 404-2). These forms are normally in a transparent plastic covered holder on a side panel, or on a seat back.

B. PRE-INSPECTION RUN-UP:

Before the airplane is actually brought into the hangar for the inspection, it is run-up and the following items checked:

1. Engine temperatures and pressures.
2. Static RPM.
3. Magneto drop.
4. Any unusual engine noises.
5. Engine response to power changes.
6. Propeller response to pitch change.
7. Proper fuel feed from each tank.
8. Idle RPM and idle mixture.
9. Alternator or generator output.
10. Instrument suction.
11. Fuel flow indication for both idle and full power.

After the engine run-up has been completed, the instrument readings are recorded and notes made of any discrepancies found, the airplane is brought into the shop and the inspection itself begun.

The inspector will most probably conduct the inspection according to systems, as this is one of the surest ways to cover every detail. We will observe this inspection in this manner.

C. ENGINE SECTION:

1. Compression check:

Probably the one check which tells the technician the most about the internal condition of the cylinders is the compression check.

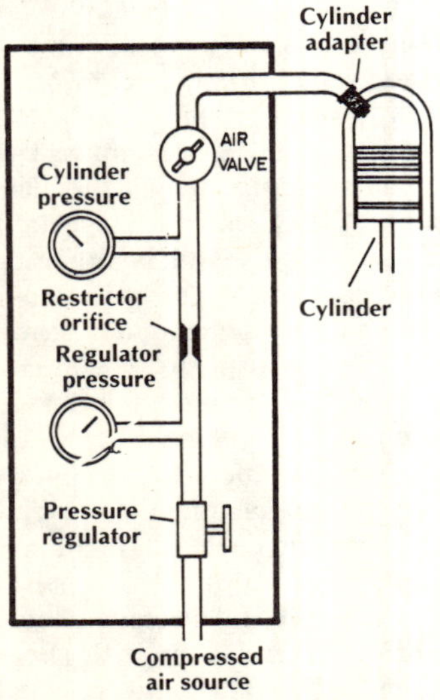

A differential compression test determines the condition of a cylinder by the amount of air that leaks past the rings and valves. The pressure differential between the two gages indicates the amount of leakage.

- Figure 93 -

While the engine is still hot from its run-up, the top cowling is removed and the upper spark plugs taken out. A differential compression tester, Figure 93, consisting of an air pressure regulator, a calibrated orifice, and two pressure gages is connected into the cylinder through the spark plug hole. The piston of the cylinder being checked is placed on top center, compresssion stroke, and the regulator adjusted to exactly 80 psi on the regulated pressure gage. Any leakage of air from the cylinder causes a flow through the tester and the cylinder pressure gage will read lower by the amount of leakage. A cylinder may be considered suspect if the leakage is more than twenty-five percent. This would show up as a reading of 60/80. One main advantage of this test is the fact that it not only shows if trouble is present, but locates the trouble. If the intake valve is leaking, for instance, the noise of escaping air will be heard through the carburetor air inlet. If the exhaust valve is faulty, air will escape through the exhaust system, and if there is piston ring blow-by, air will be heard at the oil filler or crankcase breather.

While compression tests are of value in themselves, their main use is for comparison. A difference between cylinders is of more concern to the technician than a uniformly low reading of all cylinders.

2. Lubrication system:

The oil is drained while it is still warm, and while it is draining, the screen is pulled and examined. If the engine is equipped with a full flow oil filter, it is cut open and examined for indication of metal particles or chips of any kind. New and freshly overhauled engines will most probably have small shavings or tiny particles of metal in their filters. These come from moving parts as they fit themselves together, and are of no consequence. After the oil has drained, the drain plug is replaced and safetied. The oil screen or a new filter is installed and safetied. New oil is put into the engine and the dip stick checked for indication of the proper amount.

The entire engine is inspected for oil leaks. The oil cooler, and all connecting lines, the oil temperature bulb, oil pressure relief valve, valve rocker boxes, propeller shaft oil seals, and push rod tubes are all subject to oil leaks and are carefully inspected. After the engine is looked over carefully, it is washed down with solvent and dried with compressed air.

3. Engine fuel system:

The fuel selector valve is turned off and the main fuel strainer is cleaned. This is normally a fine mesh wire screen and is cleaned by removing it, washing it in varsol or Stoddard solvent, then blowing it dry with compressed air. It is reinstalled with new gaskets, checked for leakage, and safetied. If the aircraft has an electric fuel pump with a screen, this is cleaned in the same way as the main screen. The strainer in the carburetor or fuel injection unit is removed and cleaned. Float carburetors normally have the float bowl drained and flushed out.

All carburetor controls are checked to be sure they have full travel, and the stop on the engine unit is reached slightly before the stop in the cockpit. The throttle shaft and the ends of the throttle linkage are lubricated with a drop or so of engine oil. All control wires and linkages are checked for security and freedom.

After all strainers have been cleaned and replaced, the boost pump is checked on both low and high speed. The pressure is checked on the pressure gauge and the entire engine fuel system examined for leaks.

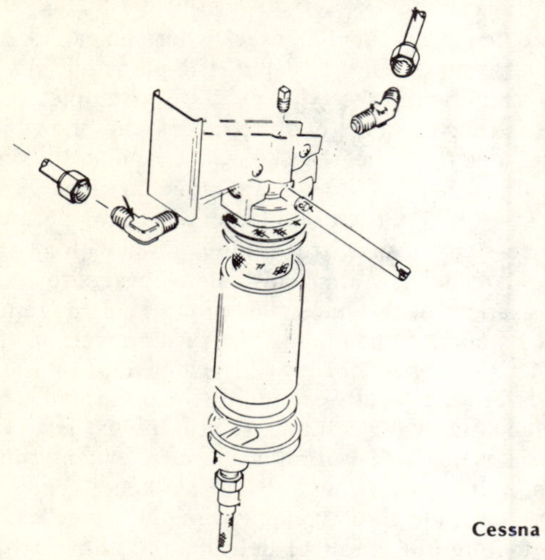

Main fuel strainers are normally fine mesh wire screens which may be cleaned by washing in solvent and blowing dry with compressed air.

- Figure 94 -

4. Induction system:

All of the air the engine breathes must come through the induction system. At the inlet of this system is the main air filter. Older filters were made of screen wire covered by a fuzzy substance called flock. These filters were serviced by washing them in a solvent such as varsol and then soaking them in a mixture of engine oil and solvent and allowing them to drip dry. Oil adhering to the flock trapped dust or contaminants trying to enter the engine. Later engines use a paper filter element similar to that used in automobiles. These filters should be checked at least every one hundred hours, and more often if operation is in a dusty area. They may be cleaned by blowing out with compressed air in a direction opposite normal air flow, and some types may be washed in warm water and household detergent, then dried with compressed air.

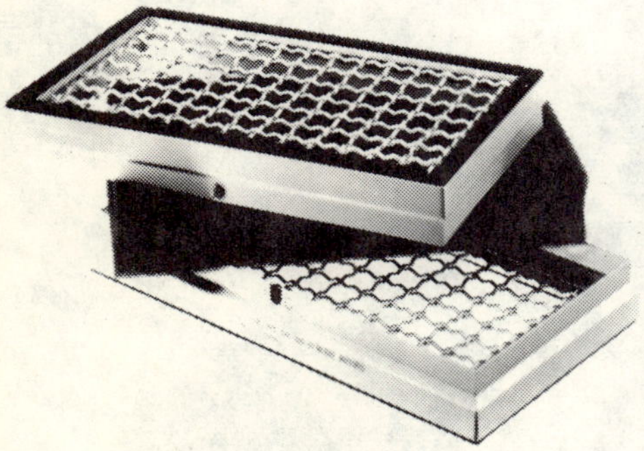

Carburetor air filters must be cleaned according to the manufacturer's recommendations.

- Figure 95 -

One of the newer air filter elements is a polyurethane foam element impregnated with a glycol chemical which attracts and holds dust. When the element is dirty, it is simply removed from its sheet metal frame and thrown away.

The importance of adequate air filtering cannot be overemphasized. Tiny particles of sand entering the engine act as a very effective abrasive, wearing away the piston rings and damaging the cylinder walls. They get into the oil system and cause abnormal wear to the bearings and internal engine parts. The silica of which the sand is composed can glaze the spark plug insulators and cause plug fouling.

Float type carburetors are very efficient mechanical refrigerators. The fuel changes from a liquid to a vapor in the throat of the carburetor, and in so doing removes heat from the air and the metal. A drop of as much as 70 degrees Fahrenheit may take place here. To prevent ice forming and choking off the air to the engine, some provision must be made to warm the air. Air is taken in from the front of the engine and passed through a shroud around the muffler, then into the carburetor. In the inspection, the A&P checks the carburetor heat control to be sure the door opens and closes fully. When the heat control is off, the heated air is directed overboard. Aircraft having pressure carburetors or fuel injection systems which are less susceptible to carburetor ice have alternate air inlets to allow warmed air from inside the engine compartment to be taken into the fuel metering system in case the air filter becomes iced over. Some systems are manually controlled by the pilot, and others open automatically if the pressure behind the filter drops because of a clogged element. These systems are checked for freedom of movement and operation.

5. Exhaust systems:

The heat energy in the fuel is only partially used to produce power; most of it is lost overboard through the exhaust system, whose exhaust stacks bolt to the cylinder heads and carry the hot gases. The pipes from the individual cylinders normally weld into a common collector and exhaust through a muffler or tailpipe. Shrouds of sheet aluminum cover parts of the exhaust system to provide heat for the carburetor and the cabin. In the

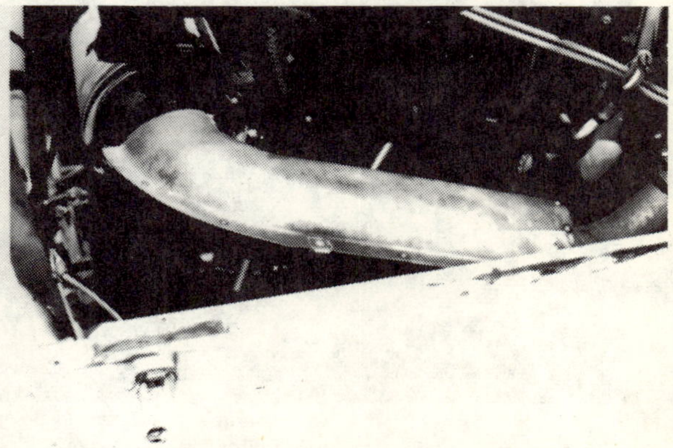

Heat for the cabin and the carburetor comes from around the exhaust pipes or muffler.

- Figure 96 -

process of inspection, all of the shrouds are removed and the entire exhaust system carefully checked for cracks or holes burned through the thin pipes. Exhaust leaks cannot be tolerated because of the danger of fire or carbon monoxide poisoning.

Exhaust gas temperature probes may be installed in the exhaust pipes, and the area around these probes is carefully checked for indication of leakage where the probes enter the stacks. Any leakage here will damage not only the stack but the probe as well.

The heater shrouds connect to the cabin heater valve with the carburetor heat box by fiberglass hose reinforced with a spiral of stiff, steel wire. When the A&P inspects this, he must make sure that it is not only in good condition, but that only hose suitable for hot air is used on the hot side of the system.

QUESTIONS:

105. Is it legal for a private pilot to conduct a 100 hour inspection on his airplane under the supervision of an A&P mechanic?
106. Is it required that a check list be followed in conducting a 100 hour inspection?
107. What check does a technician make to tell the condition of piston rings and valves?
108. What is the significance of small metal shavings in the oil filter of a freshly overhauled engine?

---

6. Ignition systems:

The ignition system is the most demanding system in an engine and has perhaps the most components to be checked.

a. Spark plugs:

When the A&P removes the spark plugs from the engine, he places them in a rack where he can identify the cylinder from which they came. The plugs tell him a story of the condition of the cylinder. If, for example, one of the plugs is much whiter than the others and shows signs of having been overheated, he will examine that cylinder carefully for indication of an induction leak. The plug shows that that particular cylinder has been operating with an excessively lean mixture. Plugs filled with a hard, gray cinder-like deposit are lead-fouled because of poor fuel vaporization, or, perhaps, the wrong heat range plug has been installed. If the plug is black with a wet, oily deposit, the A&P will check for broken piston rings or worn valve guides. After carefully examining the plugs and noting any problems, he will clean and gap the plugs. Lead deposits are broken out with a vibrator type cleaner, and the plug is given a light blasting with glass beads or a special aluminum oxide abrasive. Fine wire plugs give less trouble from fouling but are more delicate and allow only a very slight amount of blasting. After cleaning, the plugs are gapped. In this process, the plug is placed in a special fixture and the outside electrode moved over to a specified

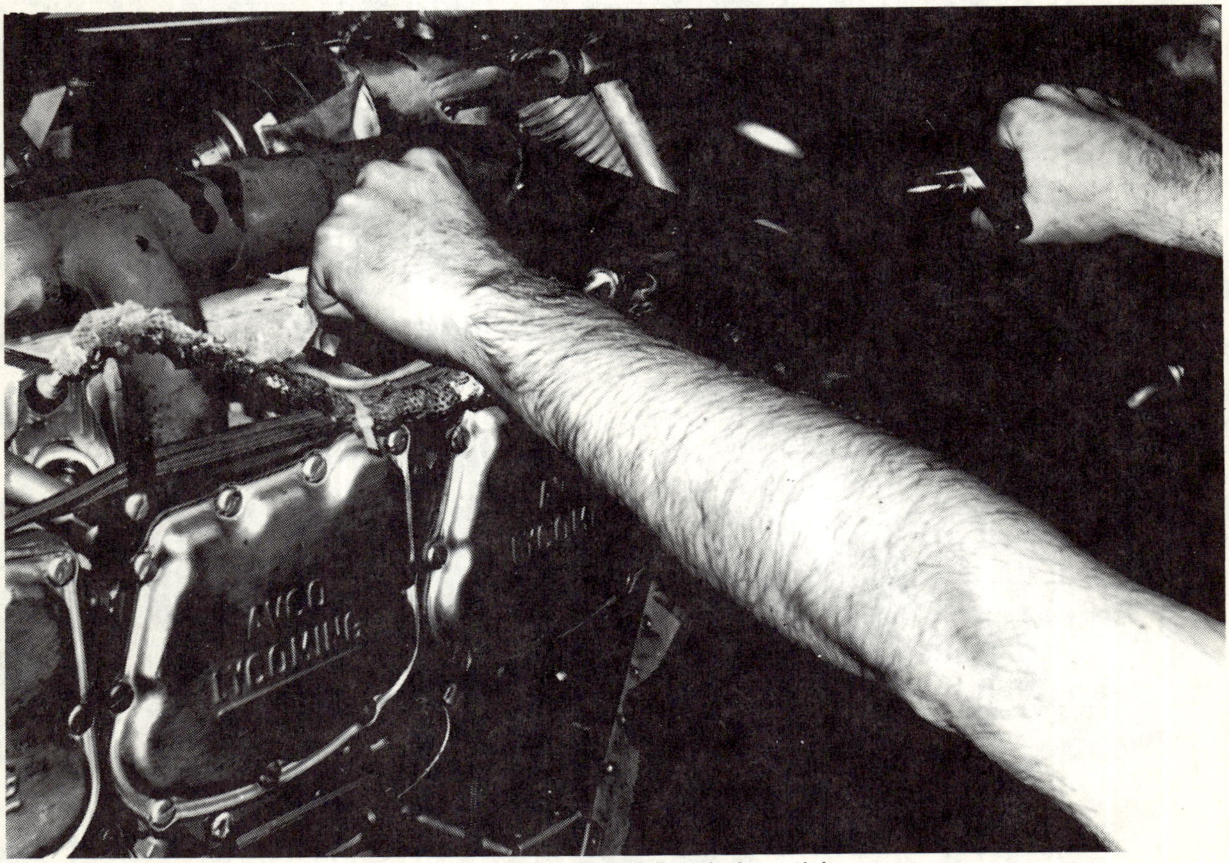

A torque wrench should be used to install spark plugs. If they are too tight, there is a danger of damaging the plug; if too loose, there is danger of the hot gases leaking.

- Figure 97 -

distance from the center electrode. The plugs are then tested and reinstalled in the engine.

The spark plug holes must be clean and there should be no burrs or carbon in the threads. A new gasket is installed and the plug torqued to the value specified by the engine manufacturer.

b. Leads:

The conductor from the magneto to the spark plug is made of a very special high voltage, shielded wire with shielded terminals at both ends. If there has been any ignition problem, or if the wire appears to be in a bad condition, a test with a high voltage leakage tester will be performed. When the "cigarette" is reinstalled in the plug, it is wiped free of any fingerprints with a special solvent to prevent the high voltage traveling across any moisture the print will attract, to ground.

c. Magnetos:

When the A&P gets to this part of the system, he removes the covers and checks the inside of the case for any indications of loose parts, damage, or oil. Oil here would indicate a failed oil seal which would have to be replaced. A magneto is a high voltage AC generator with a set of points timed to interrupt the flow of primary current at a specific time with regard to the position of the rotating magnet, and also the position of the piston in the cylinder. The condition of the points and the maximum amount they open are checked. The engine is rotated until number one cylinder is nearly ready to fire and a timing light is placed across the breaker points. A special protractor measures the angle between the neutral position of the magnet and the position at which the points open. This is called the "E" gap and its accuracy determines the intensity of the spark.

The distributor is a black plastic cap in the back of the magneto housing. A carbon brush makes contact with the rotor. As the engine turns, the distributor finger aligns with the contacts and the spark is fed to the proper plug. The distributor is carefully checked for cracks or carbon tracks; either will conduct the high voltage electricity to ground or to the wrong cylinder.

The ignition switch is checked to be sure it forms a good ground when it is in the "Off" position.

7. Electrical system:

The powerplant electrical system is responsible for generating and distributing all of the electricity needed for the operation of the airplane. This electrical energy may be produced by either an alternator or a generator. Both units accomplish the same end result, but the inspection required of them is different.

a. Alternator:

This differs from a generator in that the high current is produced in the fixed winding and does not require heavy brushes to carry it. The rectifiers are solid state diodes. Alternators are relatively trouble free, but special attention is paid to the mounting brackets, the pulley, and the belt tension.

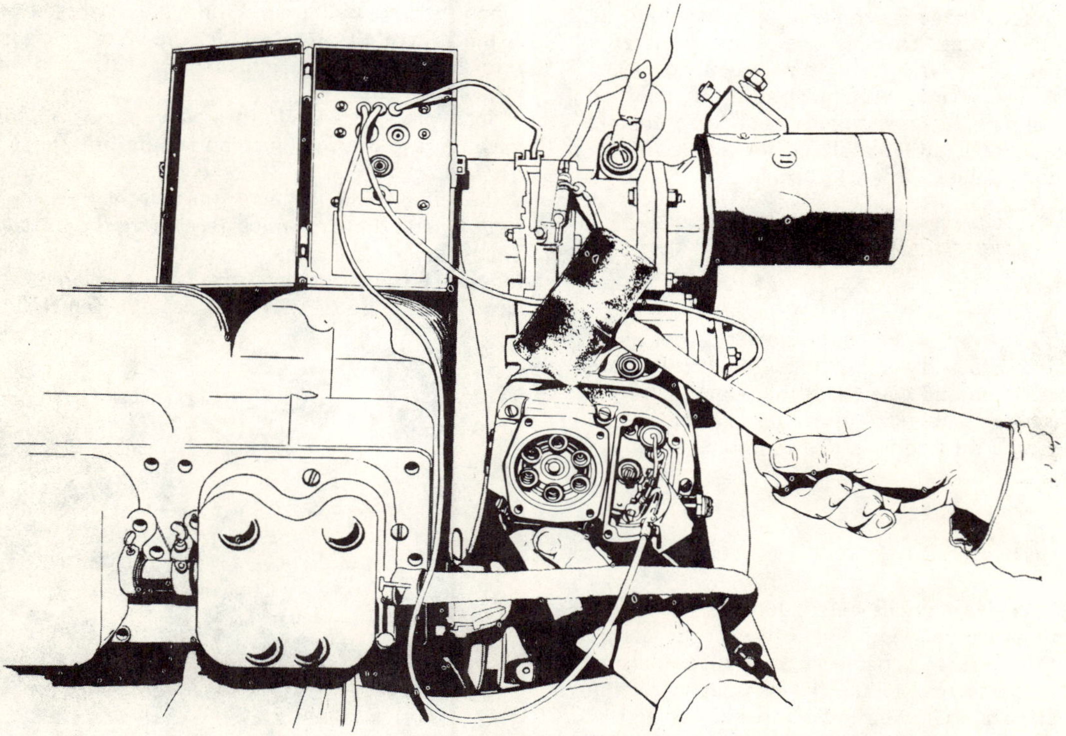

The spark at the plug occurs the instant the magneto breaker points open.
A dual timing light allows the A & P to synchronize magnetos.

- Figure 98 -

b. Generators:

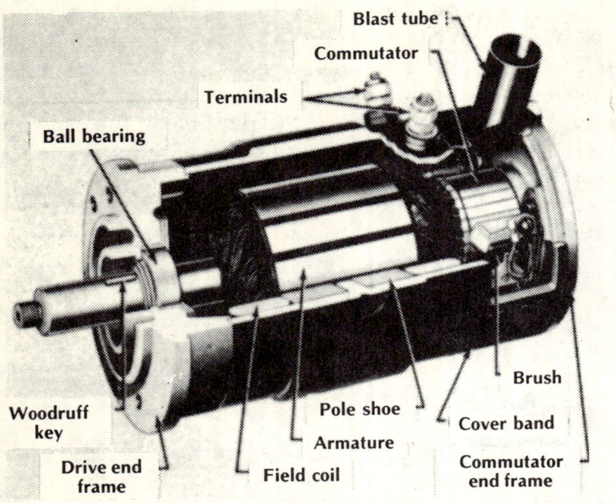

High current and long hours of use wear away the brushes and erode the copper commutator bars. The A & P must determine their condition on an inspection.

Delco-Remy

- Figure 99 -

In addition to an examination of the generator for physical condition, its brushes and commutator are checked. Brushes are subject to a great deal of wear, as are the copper bars of the commutator. The wiring and brush rigging are inspected for condition, brush tension, and for indication of overheating.

c. Voltage regulators:

Generator voltage is regulated by controlling the amount of field current. This may be done by either a vibrating point or a transistor type regulator. The generator voltage is measured with a fully charged battery in the circuit. If the voltage is outside acceptable limits, the regulator is either replaced, or removed from the airplane and adjusted on a test bench.

d. Starter:

The high current used by the starter causes much wear on the brushes and commutator. The brushes are examined for tension and wear, and the commutator for pitting or burning. The lead from the starter to the solenoid is checked for tightness and condition.

8. Cooling system:

Small air cooled engines have adequate cooling by their cylinders protruding into the breeze, but higher powered aircraft engines depend on pressure cooling to keep their temperatures within the normal operating range. The baffles are checked to be sure they fit tightly around the cylinders and the seal between the upper and lower engine compartments fits tight in the cowling. All of the cylinders are probed for cracked or broken fins and for obstructions to the air flow. Burned paint on any of the cylinders could indicate a local hot-spot which may be the result of detonation.

The cowl flaps are checked for full travel and for integrity.

9. Propeller:

The propeller blades are carefully looked over for any cracks or nicks. Running an engine up in an area of loose gravel can nick a blade so badly that there is danger of slinging a portion of the blade off. One of the most important parts of a propeller inspection is a careful examination of the blade. Small nicks and scratches are carefully dressed out with a fine file or stone. Many new propellers have square blade tips. One reason for this is to allow some tip damage to be removed without decreasing the diameter of the propeller. Most propellers are anodized. This is a treatment in which a hard, non-porous aluminum oxide film is formed on the blades to protect the metal from corrosion.

Tips are painted a bright color to make them more visible as they rotate, and the face, the side nearest the pilot, is painted a nonglossy black to keep down reflections. Any finish which has been damaged will be restored in the inspection.

The blades are checked in the hub for tightness and indications of blade wear. The track is observed to be sure the blades follow each other in their plane of rotation. Counterweights, if used, and all of the other pitch change mechanisms are checked for integrity and security.

The hub is examined for oil leakage which might indicate damaged seals or perhaps cracks. Corrosion cannot be tolerated around the highly stressed propeller, so this is carefully watched for. The bolts which attach the propeller to the crankshaft flange are checked for torque and proper safety.

Some propellers are lubricated by engine oil, but others have grease fittings for lubricating the pitch change mechanism.

Propeller spinners have grown from the small "skull cap" type to the large ones used on most modern aircraft. The spinner serves as an air flow director, and is a required piece of equipment on many airplanes today. The large size of the spinner and the high centrifugal forces

Propeller spinners and back plates are subject to large stresses and must be carefully inspected to discover any cracks before they cause failure of the spinner.

Beech

- Figure 100 -

generated as it spins place much stress on the backing plate. When the spinner has been removed, this plate is carefully checked for cracks and is replaced if any are found.

Aircraft engines with constant speed propellers maintain the RPM set by the governor. This governor is inspected for security of mounting, freedom, and proper travel of the control.

10. Accessories and general condition:

After all of the systems in the engine compartment have been gone over thoroughly, accessories such as the hydraulic pump, tachometer generator, and vacuum pump with its oil separator and relief valve, are inspected for indications of looseness, oil leaks, or improper safety. The firewall is looked over for proper sealing at any place a wire or tube passes through. All of the cowling fasteners are checked for security of attachment and the entire cowling examined for cracks.

Rubber engine mounts are checked for condition and proper torque of the bolts. The return current from the starter is high enough that special provisions must be made to prevent its flowing through the fuel and primer lines, or the engine controls. A heavy strap connects the airframe and the engine across the rubber engine mount. This carries the heavy current and minimizes the danger of fire.

QUESTIONS:

109. What can a spark plug tell about the condition of a cylinder?
110. Why must spark plug cigarettes be wiped free of any fingerprints?
111. How is generator and alternator voltage regulated?
112. What may be indicated by burned paint on a cylinder?
113. What is normally done to a cracked propeller back plate?

D. FUSELAGE AND EMPENNAGE:

1. External skin:

It was explained earlier that modern aircraft use a stressed skin type construction. This is carefully inspected for any dents, cracks, loose rivets, or missing screws. Drain holes in the lowest part of the structure are opened. The light weight structure and the highly reactive metals used in light aircraft construction make the recognition of corrosion vital, and the A&P uses every skill at his disposal to locate and treat any trace of this serious condition.

2. Internal structure:

While stressed skin aircraft do not have a great deal of internal structure, that which there is carries large loads and must be carefully inspected. Areas prone to damage from vibration such as tail posts and door frames are checked with meticulous care. It is here that the Inspection Aids put out by the FAA are especially helpful. If other technicians have found cracks or damages in the same type of airplane, they have probably been reported. This helps warn of potential trouble areas.

The landing gear box is gone over with special care, as hard landings are most likely to cause damage in this area.

3. Horizontal tail surfaces:

The horizontal surfaces are examined, as are any other structural components for indications of damage or corrosion. Specifically, the trim mechanism, whether tabs or jack screws, is checked. Attention must be paid to stabilator counterweights and their brackets in order to locate any cracks, and all control cable attachments must be inspected for freedom of movement and for the presence of rust or any corrosion. Control travel is checked against that specified in the Type Certificate Data Sheets.

If the airplane has been left out in a wind storm without the controls being locked, damage to the hinges and stops may have occurred due to violent whipping.

4. Vertical surfaces:

The vertical control surfaces are carefully gone over for condition, freedom of movement, and amount of travel, and the hinges are carefully checked and lubricated.

5. Control cables:

Some of the more simple parts of an airplane, and those whose importance is not appreciated, are the control cables. All turnbuckles* and terminals must be examined for evidence of cable slippage, and for proper safety. All pulleys must turn freely and the cables must be centered in the grooves. Cable guards must be in place over all

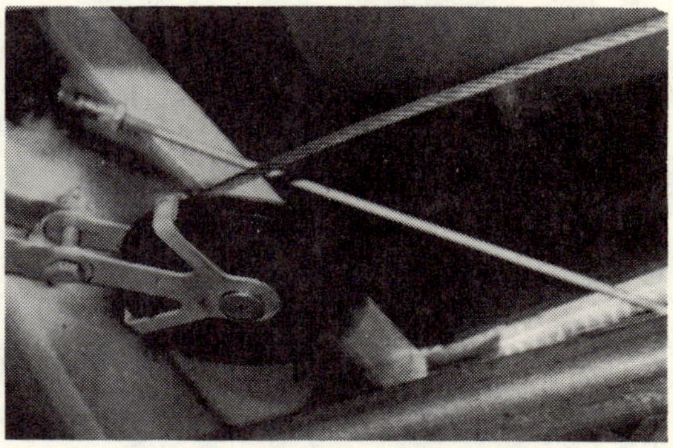

Control cables ride in the grooves of the pulleys and are held in place by pulley guards.

- Figure 101 -

pulleys and the fairleads* must not excessively deflect the cable. The cable itself is inspected for broken or worn strands and for any traces of corrosion. Cable tension is determined to be sure there is no danger of the cable jumping out of the groove, or of its being too tight for smooth operation. A special cable tensiometer* is used to measure the cable tension.

6. Seats:

Perhaps few things could be more disconcerting than to have your seat slide back just about the time you break ground. With this in mind, the seat, rails, locks, and belt attachments are carefully checked on every inspection. The FAA requires a pull test for questionable seat belts, but to be practical, if a belt is questionable, it is more economical to go ahead and replace it with a new belt than to test it. Automobile seat belts may look the same, and the difference in price may tempt you to want to use them, but no technician can pass an airplane on its inspection unless approved aircraft seat belts are installed.

7. Doors and windows:

Door locks and window latches are all checked to be sure they cannot pop open in flight, and the condition of all windows and the windshield is determined.

E. WING:

1. Skin and external condition:

The upper surface of the wing is generally operating with a compressive load in the skin. Any damage from an overload condition will show up as wrinkles in the skin and popped rivets. These indications will alert the A&P to give the structure a specially careful scrutiny.

Red or green stains under the fuel tanks or a strong gasoline odor in the wing root are indications that all may not be right with the fuel tanks.

Wing skins are thin, and there are places near the trailing edge where moisture can collect and cause corrosion; so a very careful inspection is made of every skin lap joint for traces of the white powder that indicates corrosion.

2. Ailerons:

The ailerons, like the movable controls on the tail, are checked for structural condition and for looseness in the hinges. Ailerons are usually cable operated from the control wheel to a bell crank in the wing, and then to a push-pull rod which actually moves the surface. The tension and condition of the cable is checked as well as the travel of the control. All hinge brackets are examined for cracks and the hinges lubricated.

This tensiometer determines the tension on the control cables. If the cable is too tight, it causes the controls to be stiff; if it is too loose, there is danger of the cable jumping off the pulleys.

- Figure 102 -

3. Flaps:

Wing flap hinges or tracks are checked for cracks or indication of damage, and are lubricated with the proper grease or oil. Electrical or hydraulic actuators are checked for proper operation. Flap position indicators are synchronized with the control.

The indicator beside the flap control tells the pilot the amount the flaps have extended.

- Figure 103 -

4. Wing attachment:

The attachment bolts for cantilever wings are torqued when the wing is installed, and a paint mark put across the bolt end and nut to show this has been done, and to indicate if the nut has turned since it was torqued. All of the center section structure is carefully inspected, and if the airplane uses wing struts, the attachments at both the fuselage and in the wing are checked for evidence of relative movement or looseness.

F. LANDING GEAR:

The airplane is jacked up for the inspection of the landing gear, and if the gear is retractable, a retraction test will be performed.

1. Shock struts:

With the airplane on the jacks and the struts extended, the landing gear is examined for any indications of looseness in its mounting or for any damage. If the strut is of the oleo type, it may be deflated and checked for the proper amount of fluid. Torsion links are subject to a tremendous beating, and they are checked for wear, cracks, or looseness. The nose gear is normally steerable and may develop shimmy if allowed to get loose, so particular attention is paid to this. The shimmy damper is looked over and serviced.

The amount of air in the oleo strut is determined by the amount the piston extends when the weight of the airplane is on the gear.

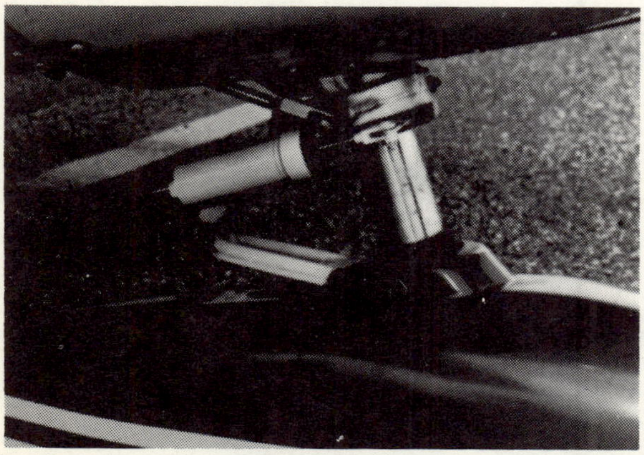

The shimmy damper prevents nose wheel shimmy by providing a hydraulic cushion between the nose wheel fork and the airframe.

- Figure 104 -

2. Wheels:

While the airplane is on the jacks, the wheels are removed and inspected, and the bearings cleaned and repacked with the proper grease. The nose wheel fork and the main gear axles are checked for any possibility of cracks or damage. The A&P will give the wheels a very careful inspection, as these are usually an aluminum or magnesium casting, and though light weight, are subject to heavy loads, especially from side loads imposed by cross wind landings. If there is any suspicion of damage, the tire will be removed and the wheel dye checked for cracks. Wheel through bolts are checked for proper torque.

3. Tires:

Aircraft tires are subject to high stresses on the tread rubber from landing loads; so the condition of the tread is carefully determined and the tire replaced if the damage is severe. The tread normally wears off of a tire before the carcass is damaged, so it is permissable to use retreaded tires. Retreading must be done, however, by an FAA approved repair station.

One problem can develop from using retreaded tires on retractable landing gear airplanes: if the retreaded tires have a larger diameter than the original, they can possibly hang up in the wells. A retraction test should be made to be sure the tire will still fit into the wheel wells with proper clearance.

4. Brakes:

On every inspection, the brakes are given a careful examination. Almost all of the smaller general aviation aircraft use single disc brakes. The disc is inspected for warpage, cracks, or pitting, and the linings checked for wear, and replaced if they show excessive use. The condition of the backing plates and housing is noted. All of the lines from the wheel unit to the master cylinder are examined for indication of leakage. The master cylinder is checked for proper operation, the parking brake for holding, and the reservoir for the proper amount and type of fluid.

5. Retraction test:

The importance of a properly operating retractable landing gear does not have to be impressed on a pilot. The comforting appearance of three green lights on the panel makes the termination of a flight a real joy.

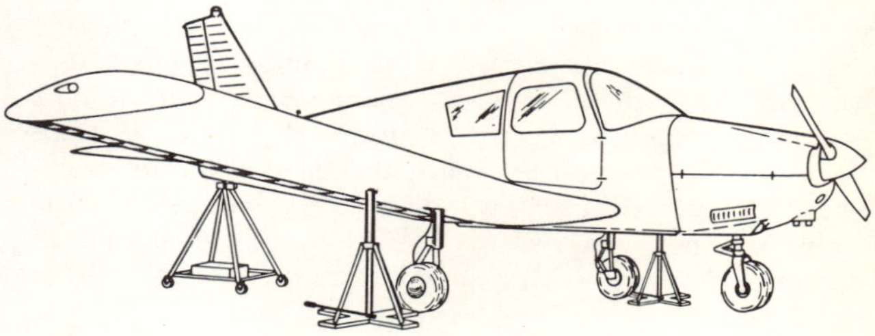

Before the landing gear can be adequately inspected, the airplane must be lifted off the floor by jacks.

- Figure 105 -

With the airplane on the jacks and all work stands clear of the gear travel, the landing gear is cycled through a retraction and extension sequence. Airplanes using electrically retracted gears usually have auxiliary power units attached to the electrical systems to conserve the aircraft battery, and hydraulically operated gear is normally operated by an auxiliary hydraulic power system.

The gear is examined for looseness, both while extended, and when it is retracted, and the freedom of movement is watched as the gear pulls up. Actuation of the down lock is noted, and the up lock, if one is used. Electric limit switches and gear position indication switches are checked for proper actuation and for security in their mounts. All of the gear doors must fit properly, and there must be no excessive play in the door hinges.

While the airplane is on the jacks and the gear retracted, the emergency extension system is inspected. When weight is on the gear, but the jacks still in place, the anti-retraction system is checked to be sure the airplane cannot be dumped on the hangar floor by an inadvertent movement of the gear handle.

G. FUEL SYSTEM:

1. Tanks:

All of the fuel tanks are carefully inspected to determine their condition. Bladder tanks are checked for the condition of the bladder and for its attachment in the structure. Fuel leaks are normally easy to find because the dye used in the fuel stains the structure anywhere there is a leak.

The filler cap is checked for proper sealing. All vent lines are checked to be sure they are clear, and all of the tank sumps are drained. The tank quantity and octane rating must be marked near the filler neck.

2. Lines:

All flexible and rigid fuel lines are carefully inspected for indications of damage or wear. Chafing can occur where the line passes through a bulkhead, or where there is relative motion between the line and any control or component.

3. Fuel quantity indicating system:

The accuracy of the fuel quantity indicating system is determined by the reading of the indicator when the tank is empty, and again when it is full. Fuel quantity systems for most light aircraft are not noted for their accuracy, so the A&P will pay special attention to the indication in the empty and near empty range. While the inspection plates are off, all of the wiring, bonding, and grounds will be examined.

4. Valves:

While the A&P is draining the sumps of the tanks, he checks the condition of the selector valve. It should move freely between all of the positions and should have a clearly defined full on and full off position.

H. ELECTRICAL SYSTEM:

1. Circuit breakers and fuses:

All of the circuit breakers are examined to see that they are in and there are no loose wires on their terminals. If the airplane uses fuses, all of them are checked to be sure the proper fuses are in the holders, and that the proper number and type of spare fuses are carried. All lights are checked for operation and condition of lenses where installed.

2. Battery:

The battery is inspected to determine the state of charge and the proper amount of electrolyte. The battery box is carefully gone over to detect any damage from spilled electrolyte. All cables and terminals must be free of corrosion and should have a coating of grease on them.

3. Wiring:

All of the wire bundles are checked for proper security, and the plugs and terminal strips examined to be sure there are no loose wires or any indication of overheating or damage to the insulation.

I. INSTRUMENTS:

1. Gyro instruments:

The filters are cleaned on the air driven gyros and on the suction regulator. The gyro instruments must indicate level when the airplane is in level flight, so the A&P will carefully note the condition of the shock mounts and of the electrical bonding straps which provide a ground return circuit for electrical gyros.

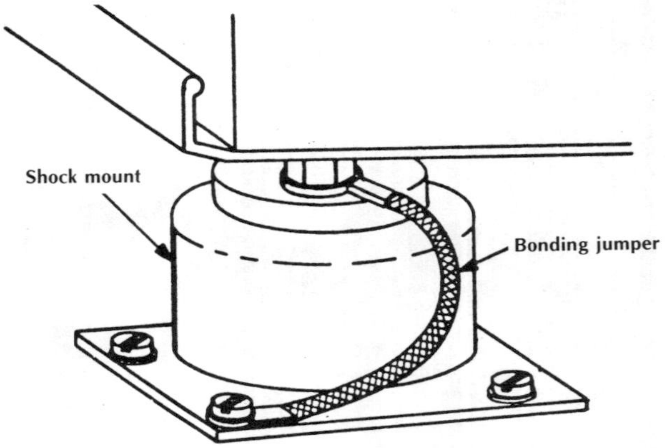

Shock-mounted electrical equipment must have a bonding jumper of sufficient size to carry the return current without an excessive voltage drop.

- Figure 106 -

2. Pitot-static system instruments:

The altimeter is the only instrument specifically required to be checked. If the airplane is to be operated under Instrument Flight Rules, the altimeter must be checked every twenty-four months for accuracy, friction, and several other characteristics. The static system must be examined for leakage and for any traces of water in the lines. Most well equipped shops can perform both of these checks with the altimeter in the airplane. A calibration card must be provided for the instrument indicating the altitude to which it has been tested.

3. Engine instruments:

On the final engine run-up, the engine instruments are observed for their proper operation and for any indications of fluctuation or oscillation.

4. Avionics:

A careful check-out of all of the navigation and communications equipment is beyond the scope of work done on an annual inspection, but on the initial run-up, the A&P

notes the operation of all the equipment; and in the course of the inspection, all of the antenna are checked for security of mounting and for the proper electrical connections. Ground wires are gone over with special care. Coaxial cable connectors are examined to be sure they make good connection and there is no corrosion forming in them.

Annual and hundred hour inspections properly conducted by an experienced A&P have perhaps more to do with extending the life of an airplane than anything else. You, the pilot, can help increase the life of the airplane and its usefulness by careful pre-flight inspections and by bringing to the attention of your A&P any discrepancy before it has the chance of growing from a small problem into a large failure.

QUESTIONS:

114. What provision is made by which technicians share information regarding potential trouble spots in an aircraft?
115. What is the best procedure to use when the condition of a seat belt is questionable?
116. Describe the appearance of corrosion in a wing skin lap joint.
117. What is the purpose of a small line of paint across the top of certain bolts and over the edge of the nut?
118. Who is authorized to retread an aircraft tire?
119. How often must the instrument static system be checked if the airplane is flown under Instrument Flight Rules?

SECTION VI: PREVENTIVE MAINTENANCE

Now that we are familiar with all the systems of the airplane and with what the technician looks for on an inspection, we can profitably look at what you the pilot can legally and safely do to the airplane you fly. F.A.R. 43.3h allows a person holding a pilot's certificate to perform certain minor preservation operations on the airplane he owns or operates - an airplane not used for air carrier service.

These items are listed in Appendix A of Part 43, but to help you perform them correctly and safely, we will elaborate on them here:

1. "Removal, installation, and repair of landing gear tires."

This little line of authorization has a lot of ramifications that must be considered. Tires, as we all know, take a real beating on an airplane. If operations are conducted from a rough surfaced runway, with lots of touch and go landing practice, the tires will have the tread worn off rapidly. On the other hand, if the airplane is operated from a sod strip and not much landing and take off practice is done, the tread will last until the carcass is worn out. Wheel fairings (pants) hide the tire and make pre-flight inspections difficult, so, for proper inspection of the tire, the fairings must be removed. The most obvious condition to look for is the condition of the tread. If the tire is worn through the tread rubber into the cord, it should be replaced. If there are cuts in the tread that extend across as much as one half of a rib, the tire should be inspected by a knowledgeable A&P. There may be superficial damage to a tire that does not render it unairworthy, but without the experience and judgment of an A&P, it is best to install a new tire if the old one is questionable.

The airplane should always be jacked from the proper jack point. Unlike an automobile, which has a sturdy steel frame, much care must be taken in jacking up an airplane properly. It is important that the right jack pads be used for this procedure - it is advisable to have an A&P show you the correct way to raise your airplane.

After the wheel is off the ground, the fairing, if one is installed, should be removed and the cotter pin taken out of the axle nut. If your airplane is equipped with Cleveland brakes, the back plate will have to be removed before the wheel can slip off of the axle. Remove the wheel and deflate the tire by removing the valve core. It is extremely important that all of the air is out of the tire before disassembling the wheel. The bead must be broken to remove the tire from the wheel, usually rather difficult to do without the use of a special bead breaker. A concentrated push must be applied to the tire right against the

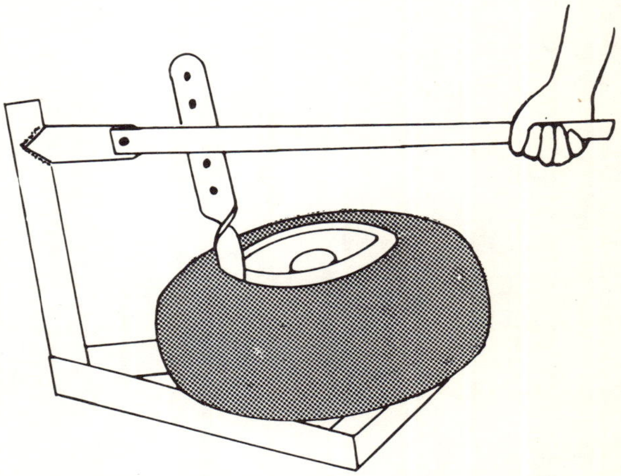

A special bead breaker is needed to loosen the bead of the tire from the rim without damage to either the tire or the wheel.

- Figure 107 -

wheel to force it loose. Don't use screw drivers or automotive tire tools to pry the bead loose, as the wheel is made of soft aluminum or magnesium and the nicks and dents these tools make will cause stress concentrations very likely to cause the wheel to fail in service. After the bead has been broken from both sides, the nuts on the through bolts are removed and the wheel taken apart. After the tire is removed, the wheel is carefully inspected for cracks, scratches, or indications of damage. If there is any questions of damage, you should get the advice of a reputable A&P. He can inspect any suspect area with dye penetrant inspection and locate cracks not visible to the naked eye.

If the tire being installed is of the tube type, the tube is carefully inspected for indications of wrinkles or pinches, and the inside of the tire lightly dusted with talcum. The tube is then slipped in the tire and inflated just enough to prevent it collapsing when the wheel is being assembled. Tires are marked with a dot on the sidewall to indicate their light point, and the tubes are marked to indicate their heaviest point. These marks should be aligned for proper balance. Slip the tire and tube over the outside half of the wheel with the valve stem protruding through the hole in the wheel half. Carefully slip the brake side of the wheel into the tire and put on the washers and nuts.

No liquid soap or lubricant of any kind should be used around the bead, and care must be taken that the tube is not pinched when assembling the wheel. After the wheel is assembled, the through bolts must be torqued to the value recommended by the manufacturer. It is important that a torque wrench be used for this operation because of the length of these bolts and the high stresses applied by a hard landing. After the bolts are torqued, the tire should be inflated to the recommended pressure, and then deflated to allow the tube to seat properly and then relax. Reinflate the tire to the proper pressure, and it is ready to remount on the airplane.

Tubeless tires are mounted in the same way as tube tires, except that there is usually an O-ring in the center of the wheel to make an airtight seal. The condition of this ring should be checked, and it should be lubricated if the aircraft service manual specifies it.

F.A.R. Part 43, in the definition of preventive maintenance, specifies that repair to landing gear tires may be undertaken by the pilot; but, practically, any damage to a tire that would require, or allow, repair would be far more profitably corrected in a repair shop and is beyond the scope of routine preventive maintenance.

2. "Servicing landing gear shock struts by adding oil, air, or both."

Like some of the other items of maintenance the pilot is allowed to perform, this is a rather simple procedure, but the equipment required makes it impractical for most pilots. Also, when a landing gear strut becomes low on fluid or air, it is an indication that some of the seals may be damaged, allowing the leakage.

Air-oil (oleo) shock absorbers must be serviced with the airplane on jacks, for the main gear, or with the nose gear held off the ground by weighting the tails down. Deflate the shock strut by depressing the core of the air valve.

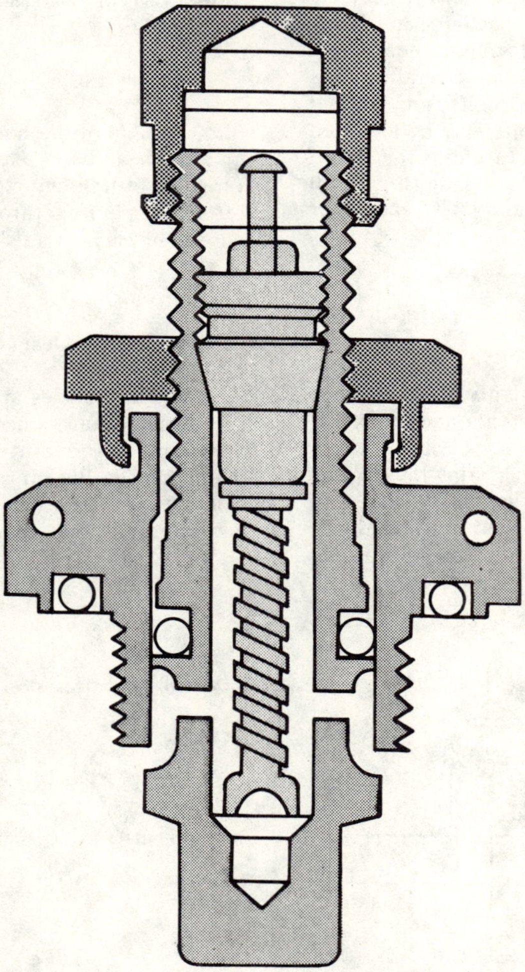

To deflate a shock strut having an AN 6287-1 high-pressure air valve, remove the cap, loosen the swivel nut, and depress the valve stem. Be careful that the high-velocity air does not strike your skin.

- Figure 108 -

NOTE: Be very careful that you do not allow the high velocity air to strike your skin, as injury can result.

When the strut is fully deflated, remove the valve core and attach a piece of plastic hose over the air valve and submerge the free end in a container of clean MIL-H-5606 hydraulic fluid. Exercise the strut several times to fill it completely with fluid and exclude any air. After the fluid has been flowing into and out of the strut, fully compress the strut, remove the plastic tubing, and replace the valve core. Some nose wheels are inflated to a specified air pressure while extended, but others, as well as all main gears, are inflated to a specific piston extension with the weight of the airplane with full fuel, on the strut. While the strut is being inflated, the airplane is rocked to prevent binding of the gear which would give an inaccurate inflation.

3. "Replacing defective safety wiring or cotter keys."

One of the trademarks of a professional aircraft technician is the neat job he does safetying nuts and bolts. This is a seemingly small thing, but its importance cannot be overlooked. The light weight of an aircraft allows vibration from the engine or propeller or jolts from landing to shake the structure enough that bolts may be loosened or nuts allowed to back off. The present trend is to use self locking hardware as much as possible, but there are still instances where craftsmanship with safety wire and cotter pins is essential.

a. Nuts:

Almost any place on the structure of an airplane where a nut and bolt are used as a fastener and the bolt is not free to turn, an elastic stop nut may be used. These nuts have a plastic or fiber insert which grips the threads and prevents the nut backing off. If the bolt is subject to turning, as in a control system, a drilled bolt should be used with a castellated nut and a cotter pin. The proper way to install a cotter pin is shown in Figure 109.

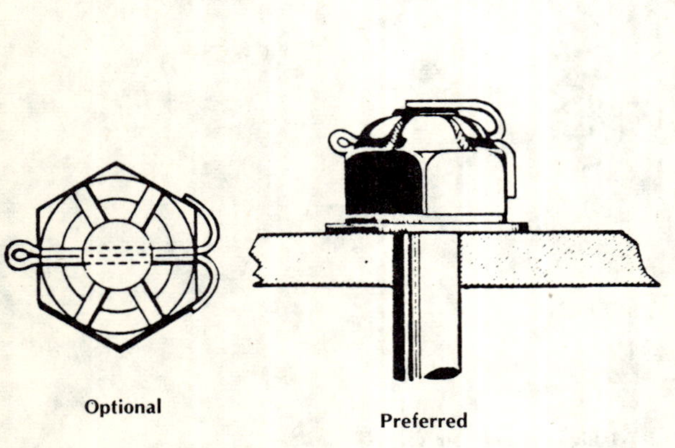

Two methods of installing a cotter pin.

- Figure 109 -

After cutting off the ends of the pin, bend them down tight against the end of the bolt and the side of the nut, so the pin is tight and has no sharp edges sticking out to cut you.

b. Bolts:

Drilled head bolts such as are used to attach most fixed pitch propellers are safetied with zinc-coated soft carbon steel, brass, or stainless steel wire. The wire is passed through the holes in the bolt head and twisted as shown in figure 110. A properly installed safety wire will

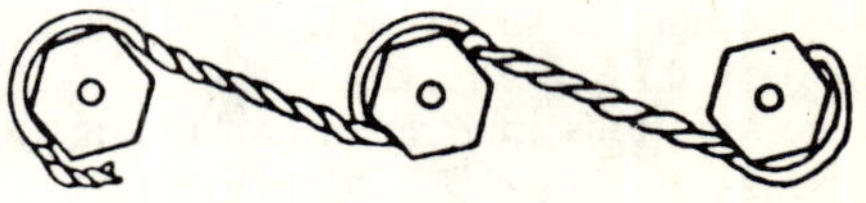

Twisted safety wire must always pull on the bolt in the direction required for tightening.

- Figure 110 -

fill about 3/4 of the hole in the bolt head, will be twisted uniformly, and will exert a tight pull on the head in the direction of tightening. The end of the wire should be cut off, leaving at least three twists, and having the end of the wire bent back so no sharp ends are exposed.

c. Turnbuckles:

Turnbuckles are the tension adjusting devices in control cables, and since this tension is critical, turnbuckle servicing is beyond the scope of pilot performed maintenance. But, if you find a turnbuckle improperly safetied on your airplane, you may replace the wire according to the method shown in Figure 111.

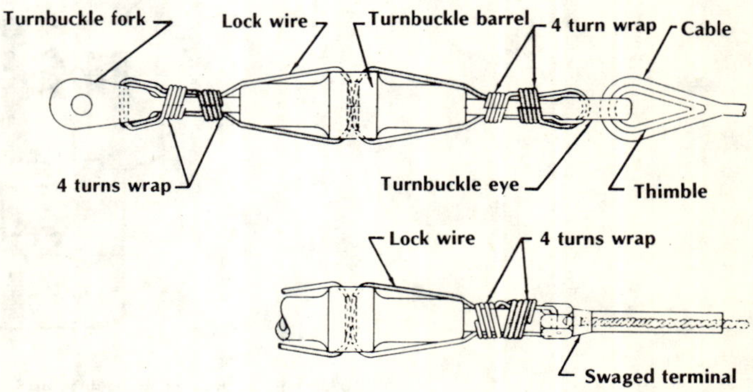

Double-wrap method of safety wiring a turnbuckle.

- Figure 111 -

Since most of the control cables in the primary system are 1/8" diameter, 0.040 inch stainless steel wire should be used for a single wrap type safetying; or, if you prefer to double wrap the turnbuckle, 0.040 brass wire may be used. Before safetying any turnbuckle, be sure the cable tension is proper. If there is any question, be sure to have it checked by an A&P.

4. **"Making simple fabric patches not requiring rib stitching or the removal of structural parts or control surfaces."**

Hangar rash, or holes from small stones or sticks, can make a fabric covered airplane really look bad. Fabric work is becoming a lost art with the advent of all metal airplanes, but with a bit of care, a patch can be applied so it will be almost unnoticeable.

Fabric aircraft may be finished with either nitrate or butyrate dope, or in some instances, have enamel over the dope. If enamel has been used, it must be removed by careful sanding in the area the patch is to be applied. Regardless of the type of dope on the fabric, butyrate dope may be used for patching. Butyrate will adhere to nitrate, but nitrate dope will not soften or stick to the butyrate finish.

Trim the damaged area to a neat circular or oblong opening and brush a coat of dope around the opening extending back at least two inches. Be very careful that the dope does not drip through the hole and onto the opposite side, as it will cause a wrinkle in that surface. After the dope has softened, scrape away the colored dope, the silver dope - the protective layer of aluminum pigmented dope - down to the clear dope. Cut your patch from new fabric of the same type on the airplane, so it will overlap the hole by two inches in all directions. Use pinking shears so the edges will stick better and have less tendency to pull up. Brush a good heavy coat of clear dope in the cleaned area and lay the patch over the hole. Pull it tight over the hole and work all of the wrinkles out while the dope is wet. Allow the dope to dry, and put a second coat over the stuck down portion of the patch. After this is dry, the entire patch should be given a couple of coats of clear dope, allowing it to dry thoroughly between coats. Sand off all of the roughness, and put on three or four coats of aluminum pigmented dope. Two coats of colored dope should be applied over the patch after the silver has thoroughly dried and been sanded with wet 400 grit sandpaper.

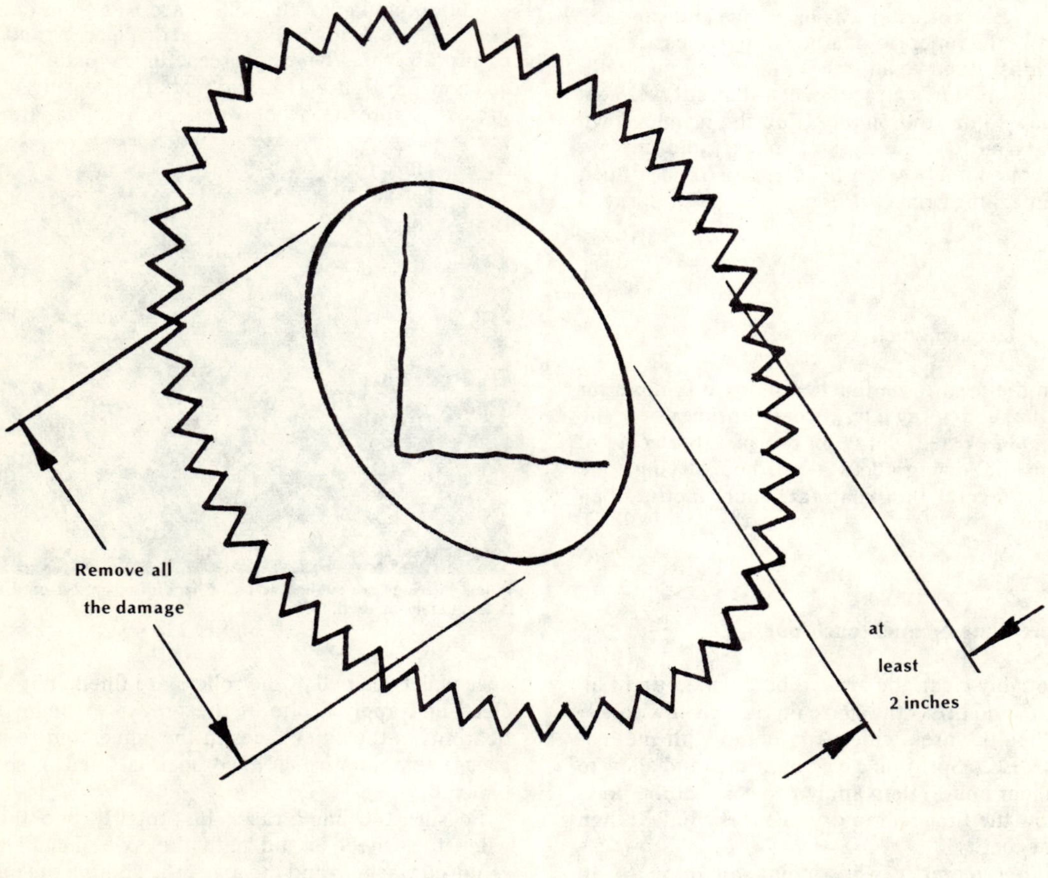

Fabric patches must lap the edge of the damage by at least two inches.

- Figure 112 -

5. **"Refinishing decorative coating of fuselage, wings, tail group surfaces [excluding balanced control surfaces], fairings, cowling, landing gear, cabin or cockpit interior when removal of disassembly of any primary structure or operating system is not required."**

While an airplane owner might want to read into this permission to repaint almost his entire airplane, an undertaking of this magnitude is clearly outside the realm of preventive maintenance. What is intended by this definition is local touch-up and spot painting. You will notice that balanced control surfaces are specifically excluded. If these surfaces are painted in such a way that weight is added behind their hinge line, they are apt to flutter.

Before attempting to do any paint touch-up on an airplane, it is a good idea to check with an A&P to determine exactly what type of finish is on your airplane. It could be enamel, lacquer, polyurethane enamel, acrylic lacquer, or epoxy, and each of these materials require a different procedure for spotting.

a. Enamel touch-up:

Scrub the area to be touched up with a solvent or thinner to remove all traces of wax or grease and sand the area with 400 sandpaper to feather the edges of any scratches or chips. Remove all traces of sanding dust with a rag damp with thinner, and spray on a thin but wet coat of primer. Allow to dry thoroughly. Rub the primer down with a clean dry piece of heavy kraft paper, and apply a light mist coat of enamel. After this dries for about fifteen or twenty minutes, the final coat of enamel can be sprayed on.

b. Acrylic lacquer touch-up:

Clean the area to be touched up as was done for enamel touch-up, and spray a light coat of primer over the area. After this has dried, spray on two or three coats of the acrylic finish, allowing about a half hour drying time between coats. Several thin coats are much better than fewer thick coats.

c. Polyurethane enamel touch-up:

Thoroughly clean the area to be touched up of all wax and grease, and break the glaze on the finish with 400 sandpaper. Wipe the area with a rag damp with methyl-ethyl-ketone (MEK). Spray on a coat of primer and allow to dry for two to four hours, then apply two coats of the polyurethane. Allow the first coat to dry to a tacky finish; then apply the final coat.

One thing to remember when doing anything to an aircraft finish: you must follow the recommendations of the paint manufacturer in detail. Do not try to modify his recommended procedure of mixing or using different primers, thinners or solvents, and do not try to apply the finish if the atmospheric conditions are not right.

QUESTIONS:

120. What kind of lubricant is used when installing a tube type tire on an aircraft wheel?
121. What type of fluid is used in an oleo shock strut?
122. Could a patch be applied to a butyrate doped surface using nitrate dope?
123. What is the best way to determine the proper primer to be used when touching up an aircraft finish?

6. **"Servicing landing gear wheel bearings such as cleaning and greasing."**

When the wheel is removed from the airplane, as it is for a tire change, the wheel bearings should be cleaned and repacked with grease. Bearings are held in their races by grease seals and seal retainers. These retainers are removed and the bearing slipped out. The bearings should be washed out in clean solvent and dried with compressed air. **NOTE:** Do not spin the bearing with compressed air, as this will cause damage.

Examine the bearing carefully for any pits or indication of brinelling. This is seen as faint impressions across the face of the bearing race made by excessive hammering or overloading of the rollers. If the races are in good condition, and all of the rollers are free from damage, the bearings are ready to be repacked. Place a good size glob of aircraft wheel bearing grease in the palm of your hand as shown in figure 113, and work the bearing into it until grease is forced out between each roller. Be sure the spaces between all of the rollers are filled. Wipe the races clean and apply a film of the grease to them. Place the bearings in the outer races in the wheel and re-install the grease retainers which have been cleaned in solvent and blown dry.

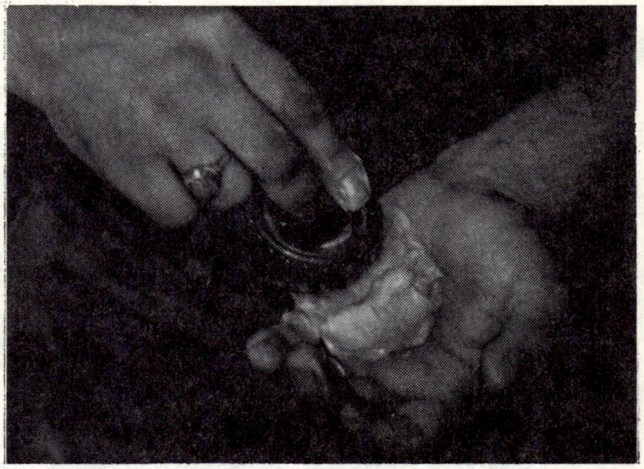

Wheel bearings are rubbed across a handful of grease until every roller is completely covered.

- Figure 113 -

Be sure the inner races fit properly into the bearing when the wheel is slid onto the axle. Install all of the required washers and the axle nut. Tighten the nut until all of the side play is out of the bearing, and then back the nut off until the first castellation lines up with the hole in the axle so that a cotter pin can be installed. Install the pin and bend its ends so that it will not foul the hub cap when it is put on.

7. **"Lubrication not requiring disassembly other than removal of non structural items such as cover plates, cowling, and fairings."**

Generally the manufacturer designs his airplane so it does not require more lubrication than the normal 100 hour or annual inspection provides. The old premise that the more grease the better simply does not apply to an airplane. Before doing any lubrication, be sure to check the lubrication chart of your particular airplane for frequency of lubrication and the type of lubricant required. Here, as with any other aspect of maintenance, if there is any question, be sure to get the counsel of a good A&P.

8. **"Replenishing hydraulic fluid in the hydraulic reservoir."**

There are three types of hydraulic fluid you may encounter: one has a castor oil base and is used for automobile brakes, not in aircraft; one has a synthetic ester base and is used in the hydraulic systems of high speed jet aircraft, not small general aviation aircraft. The fluid you will most likely use is a red, mineral base fluid known by its number MIL-H-5606. The reservoir should be checked and filled with clean fluid to the level marked. Avoid overfilling the reservoir, as it is possible for certain configurations of the landing gear and flaps to return so much fluid into the reservoir that it may be damaged.

9. **"Applying preservative or protective material to components where no disassembly of any primary structure or operating system is involved and where such coating is not prohibited or is not contrary to good practices."**

For most landplanes, this does not have much importance, but for seaplanes, operating in the hostile environment of water and spray, you are permitted to apply grease or par-al-ketone* to all of the attachment bolts and hardware of the floats, and to the control cables and pulleys.

10. **"Repairing upholstery and decorative furnishings of the cabin or cockpit interior when the repairing does not require disassembly of any primary structure or operating system or interfere with an operating system or affect primary structure of the aircraft."**

You see the cockpit of your airplane most, and this is the area where you are most likely to allow your imagination to run wild; you are on pretty safe ground, however, if you will observe some very simple ground rules.

The seats and their attachment are structural, and there should be no alteration to them; but the upholstery is not so considered, and you can work with it so long as you use material that does not support combustion, and so long as there is no possibility of your repair or alteration interfering with the controls, the seat movement, locks, or the seat belts.

This approval does not allow any modification or additions to the instrument panel, nor does it allow tape players or other music systems to be attached to the structure.

If any of the interior is repainted, it should be remembered that glare can be a problem, and nothing should be painted in such a way that it may cause distracting reflections.

Repairs to the upholstery itself can be made in the same way as repairs to household or automotive upholstery.

11. **"Replacing safety belts."**

As simple as this is, it bears a strong warning. **Do not replace the belts in your airplane with belts not specifically approved for aircraft use.** Automobile seat belts, while often looking the same as aircraft belts, may not be of the quality required for aircraft use, and their installation would not allow your airplane to pass its annual inspection.

All approved aircraft seat belts must have both parts of the belt identified by the number TSO-C-22. If this number is not clearly legible, the belt is most likely not approved. The safest way of knowing whether a belt is proper for your aircraft is to use a belt bearing the part number specified by the aircraft manufacturer. Do not alter or modify the belt attachments. Even though you may feel that your alteration is better than the original, remember, the original has been approved, and any change from this will require an FAA approval which could be quite costly and involved.

12. **"Troubleshooting and repairing broken circuits in landing light circuits."**

There are three types of landing light circuits you may encounter: retractable landing lights which lie flush with the bottom of the wing when not in use, then extend into the airstream when turned on; landing and taxi lights in the leading edge of the wing; and lights in the engine cowling. Each of these have their special problems and will be considered separately after a couple of general thoughts.

The most logical place for trouble in any lighting circuit is the bulb itself. Usually with landing lights, the clear lens allows you to see the filament and tell if it is obviously broken. If the bulb appears good, a bit of logical troubleshooting will help pin-point the problem in a minimum of time. An extremely simple tool will allow you to turn a troubleshooting problem into a simple detective job. Use almost any kind of small 12 volt bulb, such as an automotive dome light or glove box light. If you have a socket for the light, fine; otherwise, solder two wires about a foot long to the shell and the contact. With this tool the mechanics call a "hot light," figure 114, you can tell if you have voltage, and can locate an open circuit.

Check your fuses to be sure they are good; or, if circuit breakers are used, be sure they are in. Be sure the master switch and the landing or taxi light switch are both on. If you can get under the instrument panel to the back of the switches, clip your "Hot light" between one terminal of the switch and any part of the airframe structure. The light should burn. If it does, you have voltage that far; if not,

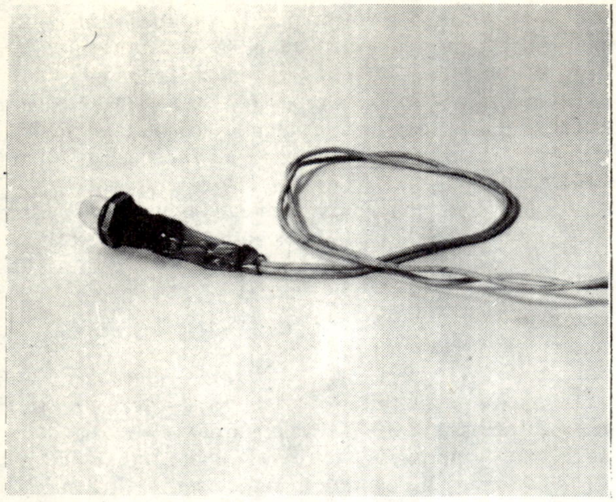

A "hot light", a 12 volt light bulb in a socket with two wires attached, is an invaluable tool for troubleshooting electrical problems.

- Figure 114 -

your trouble is in the circuit breaker or fuse. **CAUTION:** If you determine that you do not have voltage beyond the circuit breaker, it is not a good idea to probe around the circuit breakers, as the battery side of these units comprise the main bus* of the aircraft. These circuit breakers should be replaced only by an A&P. If, however, the light does show that you have voltage to both sides of the breaker, you can proceed toward the light.

a. Wing mounted lights:

Remove the cover from the landing light, or, in the case of a retractable light, lower it. There are two terminals on the back of the lamp. They should have good, clean connections. Clip one end of the test light to the wing structure and touch each of the terminals. If the bulb lights when one of the terminals is touched, you have voltage to the landing light and the trouble is with the bulb or the ground circuit. Connect the test light across both terminals of the light. If the test light burns, the landing light bulb is burned out and should be replaced. If it does not, there is an open in the ground circuit.

If there is no light when checking from the terminals of the landing light to ground, the trouble is in the wiring from the switch. An open circuit could be due to a bad quick-disconnect splice at the wing root. If an open circuit is indicated in this area, and the wiring is in a cannon plug* shared with other circuits, repair is beyond the scope of pilot performed maintenance and should be referred to an A&P.

b. Nose mounted lights:

Landing and taxi lights mounted in the nose cowl are becoming more popular all of the time, and they do lend themselves to open circuits because of vibration, and the difficulty in properly connecting the quick-disconnect splice when replacing the cowling after an inspection or engine maintenance. These wires are usually visible and can be traced out and pulled on to find the loose one. If there are no obviously loose wires, the same tests can be made as with wing mounted lights. The most likely trouble will be wire terminals broken off, and these may be replaced by crimping on a new terminal of the same type as the old one.

13. "Replacing bulbs, reflectors, and lenses of position and landing lights."

Once it has been determined that the bulb is burned out, it is usually a relatively simple matter to replace. One important word of warning, however, is the importance of using the proper part number for the bulb. The number is normally stamped on the brass body of the bulb, and the aircraft parts manual specifies the proper number for the bulbs required. When installing the bulb in its socket, you will notice on double contact bulbs, that the alignment pins are located at different heights on the sides so the bulb can be put in the socket in only one way.

When replacing landing or taxi lights which are aimed, be sure you remove only the screws which hold the lamp in the fixture and not the adjusting screws which position the beam of the light. Some landing lights are installed in the leading edge of the wing at such a point that access to them requires the removal of a wing tip. This is beyond the scope of pilot authorized maintenance and should be referred to an A&P.

Replacement of bulbs or reflectors of rotating beacons or the replacement of the bulb in anti-collision strobe lights is not included within the scope of pilot maintenance.

14. "Replacing elastic shock absorber cords on landing gear."

This is one of those short statements that hides a lot. Elastic shock absorbers are made up of hundreds of small strands of rubber, about 1/16" square. These strands are encased in a loose weave cotton braid. The older Piper aircraft, such as the Tri-pacers, the Cub series, and the Cruisers, all used this type of shock absorbers, as did the Taylorcrafts and a few others. When an aircraft of this type sits on the ramp with one wing low, it usually indicates the shock cords need replacing. There are two methods of installing the shock cords. Tri-pacers and Cubs have removable shock struts, but the Cruisers and Taylorcrafts have the cord wrapped around primary structure and must be installed in the airplane.

Changing shock cords is one of those operations in which special equipment is required. The airplane must have its nose or tail wheel, as well as the wheel on the side which is not being changed, securely blocked to prevent the airplane rolling. Jack the plane from the outboard strut attach point. When the wheel is off the ground, take out the two bolts holding the shock strut and remove the unit. Cut the old shock cord off with a knife and inspect the hydro-sorb* unit. Replace it if it does not resist in and out movement. It is far better to have an A&P install the new cords, if he has the fixture; otherwise, a chain hoist is needed to stretch the cord over the shock absorber. Be sure the cord does not roll when sliding it over the ears of the shock absorber, as this puts unnecessary strain on the rubber bands on one side.

Shock cord, as with any rubber product, will deteriorate with age, and for this reason it is coded with colored threads in the braid to indicate the year and quarter of the year in which the cord was manufactured. When buying your cords, be sure to check this. The supply house will have a chart to interpret this code.

15. "Replacing seat or seat parts with replacement parts approved for the aircraft, not involving disassembly of any primary structure or operating system."

This is self-explanatory and requires only that you adhere strictly to the parts manual for the correct replacement parts. If there is any riveting or welding to be done on the seat, this is not included in this authorization, and must be done by an A&P.

One of the critical things about the replacement of any seat is its fit in the rails and the placement of the stops. Check to be sure the seat locks completely into every adjustment hole and there is no possibility of the seat sliding backward or forward when it is supposed to be locked.

QUESTIONS:

124. How tight should the axle nut be tightened when installing an aircraft wheel?
125. What is the proper specification number for the hydraulic fluid used in most light aircraft?
126. What identification mark must seat belts have to identify them as approved aircraft seat belts?
127. How is the age of elastic shock cords determined?
128. Is an aircraft seat considered to be a structural part?

---

16. "Replacing wheels and skis where no weight and balance computations are involved."

If your airplane is approved for the installation of skis, they will be listed in the airplane equipment list, and the removal of the wheels and the installation of skis is a simple pilot-performed operation.

After the airplane has been jacked up and the wheel removed, the ski is slipped over the axle and secured with the axle nut and safetied. The safety cables are attached and the bungee to hold the nose up should be installed. Everything done in this installation must follow the instructions furnished with the skis.

17. "Replacing any cowling not requiring removal of the propeller or disconnection of flight controls."

Most modern cowling is of the two piece construction with an upper and a lower section split for removal with the propeller in place. Any piece of this cowling may be replaced with a piece bearing the identical part number.

Some cowling is attached with sheet metal screws, and these have the tendency to enlarge the holes in the cowling, so clip-on tinnerman nuts* are used to take the wear instead of the aluminum cowling.

When a cowling with cowl flaps has been removed, be sure the flap actuators are reconnected and the flap operates smoothly through its full range. Be sure the landing light is reconnected when replacing cowling having the light in the nose.

18. "Replacing prefabricated fuel lines."

Aircraft engines are subject to vibration, so all of the connections between the airframe and engine must be flexible. The fuel line between the strainer and the carburetor is a special flexible line, usually made up of medium pressure hose - not because of the pressure, but for its mechanical strength. When this line shows signs of wear, it may be replaced within the scope of your maintenance. You must, however, be sure that you use the proper fuel line assembly consisting of the hose and its end connnections. The hose connections are not lubricated, nor are they torqued specially tight as they depend on the tapered flare cone for their seal.

19. "Cleaning fuel and oil strainers."

Since aircraft not operated for hire are required to be inspected only once a year, and many fly more than a hundred hours during this period, the fuel and oil screens should be cleaned between inspections.

a. Fuel screens:

There are two fuel screens which need cleaning: the main strainer, and the screen in the carburetor. The main screen is cleaned by first turning the fuel off and removing the lower cowling. Drain the strainer bowl through the quick drain and clip the safety wire. Loosen the retainer nut and remove the filter bowl. The strainer screen should then come out. Clean it in a solvent such as naphtha or acetone and blow it dry with compressed air. Re-install the screen, using a new gasket and put the bowl back in place. Tighten the retainer as specified in the maintenance instructions and turn the fuel back on to check for leaks. If there are no leaks, safety the retainer nut in the direction the wire will tend to tighten it.

b. Carburetor screen:

Do not touch the carburetor unless you have the manual for the particular engine or have been shown exactly the proper way to do it by a competent A&P. The fuel line usually must be removed from the carburetor to get to the screen, so the fuel must be turned off. Remove the screen and wash it with clean naphtha or acetone, and blow it dry with compressed air. Re-install it, using a new gasket. After it is re-installed and the fuel line is recon-

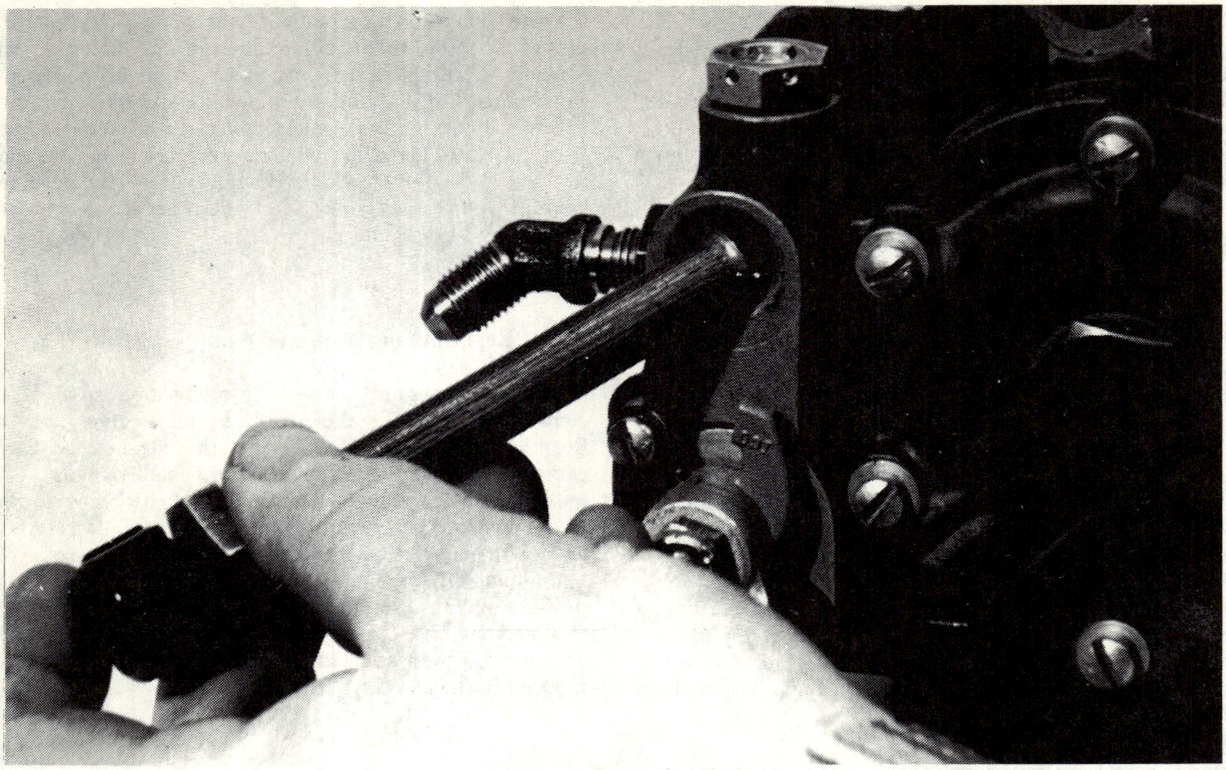

The carburetor fuel screen must be checked for leaks after it is cleaned and replaced.

- Figure 115 -

nected, turn the fuel on and the electric boost pump, if one is used, and check for leaks. If there are no leaks, the screen must be safetied in the manner shown in the instruction manual.

Fuel system screens are not complicated to service, but they must be treated with respect. Do not force any screen, and if they leak after reassembly, do not attempt to stop the leak by continued tightening. They are sealed by a gasket, and if they leak, it is almost assuredly a case of foreign matter under the gasket, or a warped housing. Remove the strainer and recheck both surfaces and the gasket. To not attempt to use an old gasket, as it will almost always leak.

c. Oil screens:

Aircraft engines may use either a simple metal screen or a paper type, full flow filter similar to that used in automobiles. As a matter of practicality, the pilot may clean the wire screen type, but without the proper special tools for removal and proper installation, it is better to have the full flow type filter changed by an A&P.

Screen type filters may be removed by first removing the oil temperature probe and then the filter. Some engines require a large open end wrench for this, while others have a small cap holding the filter in place. Some oil will drain out when the filter is removed, so you will need a small bucket or pan to catch it. When the filter is out, carefully examine it for indication of the engine condition. There may be some carbon particles if the engine is old and is using straight mineral oil, but if there are indications of metal, you should consult your A&P to find out where it most likely came from, and what should be done on the basis of this information.

If there is no contamination, the screen is washed in naphtha or varsol and dried with compressed air. Check the strainer cavity in the engine to be sure there is no foreign matter there, and re-install the screen, using a new gasket. Be very careful that there is no binding. Torque the strainer to the value specified in the maintenance manual, and safety as the manual specifies. Re-install the oil temperature bulb if it has been removed. The operation of cleaning filters is not complete until the engine has been run and checked for any indication of oil leaks.

20. "Replacing or cleaning spark plugs and setting spark plug gap clearance."

The spark plug on an aircraft engine can tell you as much about the condition of the engine as any other one thing. You as a pilot are allowed to service spark plugs, and this may be done effectively if a few simple points are observed.

If the magneto checks have been indicating increasingly larger drops as the time builds up, and the plugs foul more easily from prolonged idling, it may be a good idea to check them.

Before starting any spark plug servicing, it is essential to do some planning. You must have the proper tools for removal and replacement, and the equipment for cleaning and gapping. You will need an open end wrench of the proper size to remove the leads, and the proper size deep socket wrench to remove the plug. Sometimes these plugs are really tight in the threads and require a 1/2" drive socket and a long break-over handle. Some engines require extensions and universal joints to get to the plug. Unscrew the nut holding the lead in the plug and carefully remove the lead by pulling it **straight out** of the plug. When removing the plug, be very sure the socket is all the way down on the plug and your pull is straight, not causing the socket to bind on the plug. This will surely break it. Remove the plug and put it in a marked tray so you can identify the cylinder from which it was removed. After all of the plugs have been taken out, study them carefully. Plugs with oily insulators in the firing end indicate worn rings or valve guides. Black soot on the plugs indicates too rich a mixture. Insulators that appear too clean might indicate that the cylinder is running too hot, probably from an induction system leak.

AC

A dry, black soot in the firing end of a spark plug indicates operation with an overly rich mixture.

- Figure 116 -

If the plugs appear normal except for the usual deposits and wear, they may be cleaned. Lead deposits between the nose insulator and the shell may be carefully removed by using a hacksaw blade ground down into a long, narrow pick. Break out all of the clinker-like lead deposits and give the plug a very light abrasive blast. Do not attempt to remove the lead with the abrasive as too much of the electrodes will be worn away in the process. After the plug firing cavity is clean of lead deposits and carbon, the plug is put in a gapping fixture and the outer electrode moved over, very carefully, to the proper distance from the center electrode. The special gapping tool allows this gap to be closed evenly so the firing will not be concentrated at the outer tip of the gap. Use only a wire gauge to measure the gap so it can be accurately measured all the way across. After the plug has been cleaned and gapped, it must be checked for firing under pressure. Almost all aircraft spark plug cleaning machines have a tester with them.

The terminal cavity of the plug must be wiped out with a cloth damp with a solvent such as tri-chlor-ethylene. A small amount of anti-seize compound may be placed on the second thread from the firing end, being very careful that no compound is allowed near the electrodes. Replace the plugs in the cylinders using new gaskets and tighten them with a torque wrench to the value specified for your particular engine. After the plugs have been torqued in, re-install the leads, being very careful that the "cigarettes" go down in the plugs without binding. Before inserting the leads, inspect them for cracks and wipe any oil or fingerprints off with a cloth damp with tri-chlor-ethylene.

21. "Making small simple repairs to fairings, non-structural cover plates, cowlings, and small patches and reinforcements not changing the contour so as to interfere with proper air flow."

If you are the pilot of an older airplane, you are aware of the many cracks which develop around the engine cowling. The light weight material used for these large pieces makes cracks easy to develop, and if a crack is not properly checked, it can grow large enough to require a patch. Any time you discover a crack in a piece of cowling, or non-structural metal, drill a small hole, about 1/8" in diameter at the end of the crack. This will spread the stresses so they will not continue to tear the metal. If the crack is long enough to require a patch, a piece of aircraft aluminum the same thickness as the cowling should be cut long enough and wide enough to allow a row of rivets to be placed along both sides of the crack with the center of the rivet holes at least two rivet diameters from the edge of the crack and from the edge of the patch, figure 117. For most

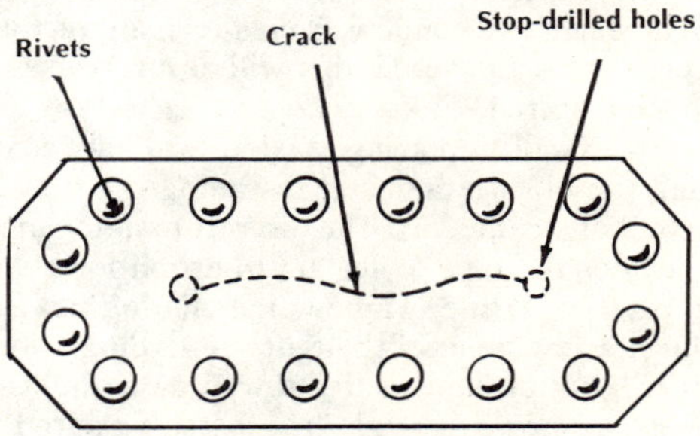

Riveted patches must use rivets having a diameter of approximately three times the thickness of the skin, and there must be a minimum of two rivet diameters between the center of the rivet hole and the edge of the crack or the patch.

- Figure 117 -

patches in cowling, 1/8" AN 470-A rivets may be used. These are made of soft aluminum and may be driven without using a rivet set. Cut the rivet just long enough to stick through the cowling and the patch by about 3/16" (1-1/2 diameters). Hold the head of the rivet against a hard, flat surface and flatten the end of the rivet with a hammer to about 1/16". If you are careful to get your rows of rivets straight, and lined up with a factory installed row of rivets, a patch such as this can be made so it will be little noticed.

Fiberglass wheel fairings and cowling are more susceptible to damage than metal, but are also more easily repaired. Remove the cowling or fairing and clean the inside of the surface around the crack thoroughly with methyl-ethyl-ketone. Fiberglass repair kits may be purchased at boat repair shops. Mix the resin and accelerator as directed by the instructions, and cut several pieces of fiberglass cloth to cover the crack. The first layer to be applied should be wide enough to cover the crack with about a half inch overlap on each side. Each succeeding layer should extend about a half inch on each side of the lower layer for proper build up. Lay four layers of cloth in the resin, working all of the air bubbles out of each layer. After all of the layers are laid up, allow the resin to cure. This can be hastened by using a heat lamp, but be sure you do not use more heat than recommended by the repair kit instructions. After the repair is made, sand the outside surface smooth and refinish it to match the rest of the cowling or fairing.

22. "Replacing side windows where that work does not interfere with the structure or any operating system such as controls, electrical equipment, etc."

Almost all transparent plastic used for aircraft windows today is an acrylic which has a relatively high strength, is easy to work, and does not discolor as some of the older plastics did. One problem with acrylic windows is their tendency to craze or form tiny surface cracks especially where the plastic is curved, as in a windshield, and when it has been exposed to extremes of temperature.

Pilots are allowed to replace side windows as these are not subject to the air loads of the windshield, and because their smaller area makes them less susceptible to damage from improper installation. It is possible to buy sheet acrylic and cut a new window using the old one as a pattern, but experience will prove that you are better off buying a ready cut window from the manufacturer or a reputable supply house, as this will assure you of having the proper material.

The process of replacing a side window consists of removing the old window and installing the new one the way the old one came out. The aircraft maintenance handbook specifies the type of sealant to use around the edges of the plastic. While window installation is a simple procedure, a few general thoughts regarding transparent plastics will not be amiss. If the window is held in place with rivets or screws through the plastic, be sure the holes are considerably larger than the fastener so any dimensional changes in the window will not create stresses which may cause cracking or crazing. One of the best ways to get a hole in an acrylic window is to melt it through, using a punch of the right size, heated in the flame of a propane or acetylene torch. A few practice holes in an old window will show you just how hot to get the punch. Sealing around the windows is done with gaskets or sealing compound. Acrylic windows are protected with paper or a plastic film to prevent scratches. Leave this on until the window is in place. The protective coating may be removed around the edge where the installation is made, but leave as much intact as practical. Any masking tape or adhesive from the protective paper may be removed by gently rubbing it with a rag saturated with alapathic naphtha.

23. "Replacing any hose connection except hydraulic connections."

Replacement of fuel hose has already been discussed, and since hydraulic hoses are specifically excluded from pilot maintenance, this leaves essentially the hoses in the instrument system.

a. Vacuum system hoses:

In the past, vacuum lines to the instruments have been regular cotton braid reinforced rubber hose with AN type fittings on the ends. Recently, airplane manufacturers have been taking advantage of transparent polyvinyl chloride tubing connected to the instruments by slipping it over beaded tube fittings. Ratchet type clamps secure the hose. Replacement is done by duplicating the hose and slipping it over the fittings. The clamps are released by slipping the point of a small screwdriver under the tongue of the lock and snapping it up. The clamp is replaced by slipping it over the hose and sliding the ratchet into position and tightening it by squeezing the bridges together with a pair of pliers.

The spring-type instrument hose clamp is adjustable by the ratchet action of its teeth.

- Figure 118 -

b. Engine instrument hoses:

Oil pressure, manifold pressure and fuel pressure gages are connected to their respective pick-up points in the engine with a low pressure rubber hose. If any hose needs replacing, get a replacement from the aircraft manufacturer bearing the correct part number. The fittings seal with the flare taper, so no lubricant is needed. When replacing any hose that runs through the firewall, be sure to seal the firewall in exactly the same way it was originally sealed. It is important that no leakage be allowed between the engine compartment and the cabin.

24. "Removing and installing glider wings and tail surfaces that are specifically designed for quick removal and installation and when such removal and installation can be accomplished by the pilot."

This book is specifically written for pilots of powered aircraft, and this authorization has no application to this type of aircraft.

25. "Replacing batteries and checking fluid level and specific gravity."

Aircraft batteries are similar to automobile batteries except for their physical size and therefore their electrical capacity. Special attention should therefore be paid to the battery in order to get the maximum service from it.

Simple battery service consists of keeping the battery clean and free from any moisture collection on its top. The electrolyte level should be kept just above the plates. Most batteries have a divider that serves as an indicator for the proper water level; if low, it should be adjusted with either distilled water, or any water sufficiently pure for drinking.

Batteries should never be left in a discharged condition. If they are, lead sulfate on the plates will harden and be almost impossible to convert back into lead and lead peroxide.

The condition of charge of a battery is measured with a hydrometer. This measures the specific gravity of the electrolyte. When the battery is fully charged, the specific gravity will be between 1.275 and 1.300. If it drops as low as 1.200, it should be recharged. Batteries should be charged by the constant current or slow charge method, if at all possible. It is possible to damage a battery with a "quick" charge if enough current is allowed to flow into it that the plates overheat and warp.

When a battery is removed from the airplane, the master switch should be off, and the ground terminal removed first. Doing this prevents a spark, if the wrench shorts the terminal to the battery box. Before replacing the battery, be sure the box is free from any corrosion. If there is any indication of corrosion, it should be carefully sanded out with plain sandpaper, not emery, and the area washed thoroughly with bicarbonate of soda and water. Rinse it and dry with compressed air. Prime the area with zinc chromate primer and paint it with an acid proof paint. When the battery is re-installed, be sure to block it in the box with the same spacers as were originally used. Replace the cables, removing any corrosion from the terminals. Connect the cable to the solenoid first, then the ground cable. A coat of vaseline or light grease helps protect the terminals from corrosion.

QUESTIONS:

129. What kind of fastener is used with a sheet metal screw to prevent the hole in a soft aluminum cowling from becoming enlarged?
130. What should you do if you find metal particles in the oil screen you are cleaning?
131. What could be indicated if the firing end of one spark plug is much cleaner than that of the other plugs taken from an engine?
132. How are lead deposits removed from a spark plug firing end?
133. What causes the tiny cracks that often appear in aircraft windshields?
134. What kind of water may be used to service an aircraft battery?
135. How is the condition of charge of an aircraft battery measured?
136. When removing a battery, which cable is removed first?

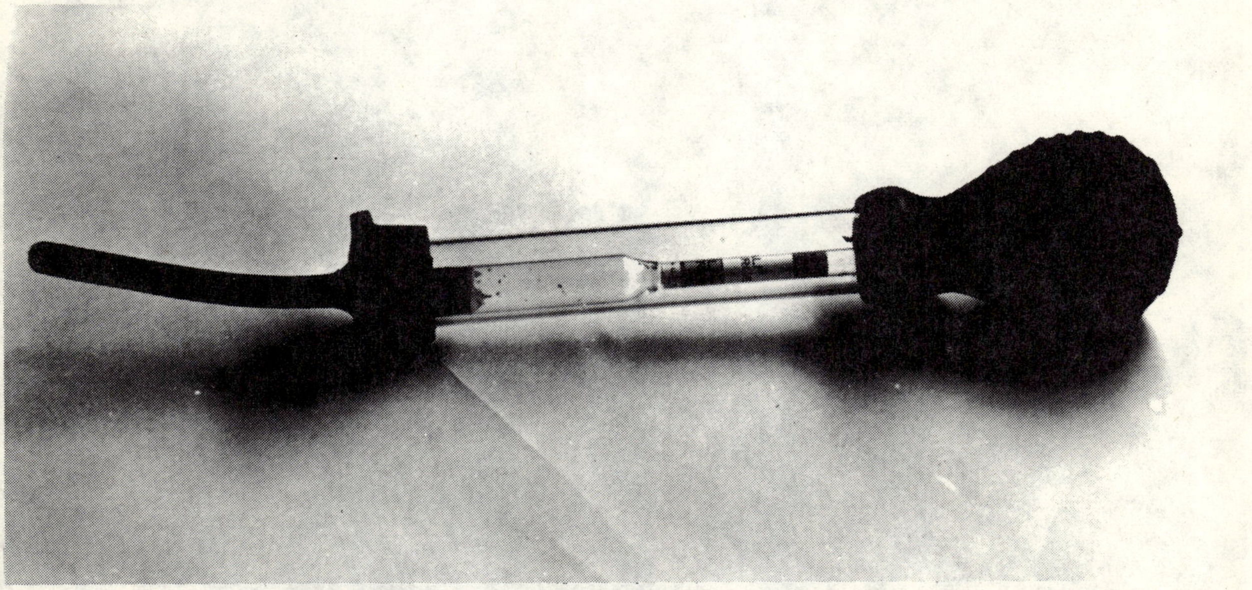

The hydrometer is used to measure the specific gravity of the electrolyte in a lead-acid battery to determine its state of charge.

- Figure 119 -

IIAC
N400M
N5401P

SECTION VII: GROUNDS RULES FOR PILOT-PERFORMED MAINTENANCE

A. RELATIONSHIP WITH THE FIXED BASE OPERATOR:

The matter of a pilot doing some of his own maintenance can be a touchy subject around some airports. In fact, on some fields it is strictly prohibited as a matter of airport policy. On others, it is tolerated; while on some it is encouraged, and provisions made for pilot-owners to become involved in the routine maintenance of their airplanes.

If pilot maintenance is allowed where you base your airplane, be sure you make arrangements with the shop manager about the use of the space where you will be working, for the use of equipment such as jacks, or for the use of shop air or the solvent spray booth.

You will have to make arrangements with the parts room to cover the myriad of small parts, such as bolts, nuts, gaskets, and cotter pins. These don't amount to much for each item, but in the aggregate they can come to an appreciable sum.

Clean-up facilities are a touchy subject on any airport. Drain oil should be properly disposed of. Do not pour it out behind the hangar if you expect to stay in the good graces of the management. Solvents used for cleaning the airplane or washing parts are usually flamable and constitute a fire hazard. Every precaution should be taken to prevent carelessness from jeopardizing your opportunity to do maintenance on the airport. Getting overspray from a cleaning gun on someone else's airplane is a good way to lose friends and influence people - for bad.

B. RELATIONSHIP WITH THE A&P TECHNICIAN:

As you become involved in doing some of your own maintenance you will need, occasionally, to consult an A&P. A good relationship with a competent technician can make working on your airplane a real pleasure, and at the same time increase your flying safety by helping you to know your machine better and by being able to detect small problems before they become large troubles. A good relationship can be developed if you will observe a few logical rules of courtesy and consideration.

The technician's knowledge and his time are his stock in trade. He has not come about his knowledge and experience without much study and hard work, so respect this and be willing to pay for his expertise.

The responsibility of the A&P is heavy, and he had much rather be doing his job, completely unhindered by the pilot. Respect this, and don't pull him off of an inspection to answer a question or in any other way interfere with his primary work.

Above all, **DON'T BORROW TOOLS**. This should go without saying, and is a tender spot with all A&P's. He has invested a sizable sum in his hand tools and specialized equipment. These, he guards jealously. You may find an A&P friendly and amiable about sharing his knowledge, but he becomes very uncooperative about sharing his tools. Respect this characteristic and the relationship will profit.

C. TOOLS REQUIRED FOR PILOT-PERFORMED MAINTENANCE:

Preventive maintenance can be performed with basic hand tools such as listed here. Special tools such as dial indicators, compression testers, timing lights, and rivet guns are normally required only for maintenance beyond the scope allowed the pilot and so should pose no problem.

As with all mechanical work, it is a good policy to not buy inferior grade tools that do not fit the hardware, tools that will slip off and damage the nuts and bolts, to say nothing of your knuckles. This list of basic tools will allow you to perform the majority of pilot-authorized maintenance.

- Combination wrenches (box and open end), 3/8'' through 3/4''.
- 7/8'' deep socket, 1/2'' drive, for spark plug service.
- 1/2'' drive break-over handle.
- 3/8'' drive socket set consisting of at least the following:
  - 3/8'' through 3/4'' sockets (6 or 12 point).
  - Universal joint.
  - 2'' and 6'' extensions.
  - Ratchet handle.
  - 3/8'' female and 1/2'' male adapter.
- 3/8'' drive torque wrench 0-600 inch pounds.
- 8'' adjustable end wrench.
- 6'' diagonal cutting pliers.
- 8'' duckbill pliers for safety wiring.
- 6'' slip jaw pliers.
- 8 ounce soft face hammer.
- 6'' slot screwdriver.
- #1 Phillips screwdriver.
- #2 Phillips screwdriver.

In addition to these tools, you will need a roll of 0.040'' stainless steel safety wire and a ''goody box'' of miscellaneous hardware such as sheet metal screws, small nuts, washers, machine screws, and cotter pins. A tray to hold spark plugs with holes for each plug is essential for spark plug servicing.

One of the most important items of maintenance equipment is a copy of the aircraft service manual for your particular airplane. This AMFI text provides background information and general procedures, but the manufacturer's service manual includes details to be considered. Specific instructions always take precedence over general information.

D. PARTS:

One extremely important item to be considered when discussing pilot-performed maintenance is the matter of parts. When the decision is made to do some of your own maintenance, a firm resolution should be made not to hedge on parts. If you work under the supervision of the mechanics in a fixed base operation, it may be part of the agreement that parts will be purchased through them. This is an understandable arrangement and one that helps keep the FBO in business to be of help to you. You will pay list prices for these parts, but the service you receive makes this a wise purchase.

There is a lively trade in bogus* parts whose price makes them wonderfully attractive to the unknowledgeable flying public. You can often find ''genuine'' manufacturer's parts for a fraction of what you would pay if you bought the ''same'' part from your local FBO. When you read advertising of this type, be cautious. Some dealers in this type of merchandise are less than scrupulous and sell repackaged military surplus materials, reclaimed salvaged parts, or new parts which have been manufactured to standards not compatible with your requirements. These parts are sold with no greater liability than the sale of any other misrepresented goods. It is when this hardware becomes part of a certificated airplane that a serious offence has been committed, and this offence is charged to the person installing the part, not the one who sold it. The recourse used by careful technicians is to be sure of the integrity of the source of his materials. For this reason, premium prices are paid for hardware and parts used by reputable fixed base operators, but the knowledge of the source of parts and the reputation of the vendor makes this a sound policy.

QUESTION:

137. Who is responsible for compromising the safety of an airplane by the use of bogus parts, the vendor of the parts, or the one who installs them?

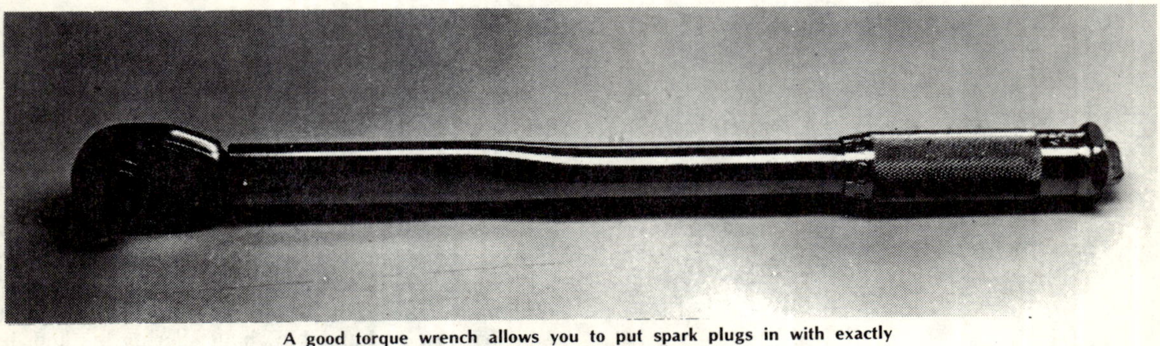

A good torque wrench allows you to put spark plugs in with exactly the torque recommended by the engine manufacturer.

- Figure 120 -

AVCO

SECTION VIII: FINE POINTS ON AIRCRAFT ENGINE OPERATION

A. CARBURETOR ICE AND THE PROPER USE OF CARBURETOR HEAT:

The National Transportation Safety Board has conducted an intensive study of aircraft accidents due to engine failure, and in an appreciable number the culprit was found to be carburetor ice. This is easy to understand when it is realized that a great amount of heat energy is required to convert liquid gasoline to gasoline vapor. This heat comes from the air and from the metal of the carburetor. When the temperature of the air drops, the amount of moisture it can hold in a vapor state decreases, and water changes into its liquid state. If the temperature drop is sufficient, and it can be as much as 70 degrees Fahrenheit, the water will be below its freezing point and it will freeze around the butterfly valve in the carburetor throat. This ice chokes out the air and disrupts the mixture produced by the carburetor to such an extent that the engine loses power, gets rough and can stop.

To prevent the formation of ice in the carburetor, the FAA requires all normally aspirated engines to be able to raise the temperature of the incoming air by 90 degrees F. This is done by routing the air through a heater muff around the exhaust and then into the carburetor.

Heating the air entering the carburetor prevents the formation of ice, but it is not without its problems. If the air is too hot when it enters the cylinders, the heat added by compression will raise the temperature to such a point that the fuel-air mixture will explode or detonate rather than burn smoothly as it should. A second danger in the excess use of carburetor heat is the fact that this air is not filtered, and if used on the ground, the air drawn into the engine through the carburetor is likely to contain sand, a very effective abrasive. Engine parts will be damaged by excessive wear and the spark plugs may be glazed over with a coating of silica from the sand.

Carburetor ice does not occur only when the outside air is in the freezing range. It can occur any time the water **inside** the carburetor is in the freezing range.

The first indication of ice with an engine equipped with a fixed pitch propeller is a drop in rpm and perhaps a trace of roughness. If the airplane is equipped with a constant speed propeller, a drop in manifold pressure signals the onset of ice. When ice is suspected, increase the throttle to restore the lost power, and apply full carburetor heat. If ice is the cause of the power loss, as soon as the ice has melted, the power will increase. If ice was not the problem, the power will stay where you have set it.

When icing conditions are encountered, the ice should be removed by the use of full carburetor heat, then the heat control returned to the cold position. Periodic application of full heat will prevent further formation of ice. The use of partial carburetor heat for ice control is not recommended unless the airplane is equipped with a carburetor mixture temperature gage. This instrument uses a probe to measure the temperature inside the carburetor and is far more useful than the carburetor air temperature gage which measures the temperature of the air as it enters the carburetor. In figure 121, we see that the coldest point in the carburetor is just downstream of the venturi where the air pressure is low. It is here that the ice normally begins to form. If partial carburetor heat is applied, the point of initial ice formation will move downstream where a larger amount of ice can accumulate before the customary warning conditions occur. By the time enough ice has formed to warn the pilot of its presence, the engine could have conceivably lost power to the extent that there will not be enough heat to melt this ice, which, because of its location, is more difficult to remove than it normally would be. If a carburetor mixture temperature indicator is installed, partial heat may be employed to maintain the mixture at a temperature of about five degrees Celsius (41 degrees Fahrenheit). This is slightly above the freezing point of water and provides the most dense condition of the inlet air, for best power production, without the danger of ice formation.

The application of carburetor heat decreases the density of the air entering the engine, and since the fuel is metered into the airstream on the basis of the volume of air, application of carburetor heat will enrich the fuel-air mixture. This characteristic may be taken advantage of when adjusting the mixture in flight.

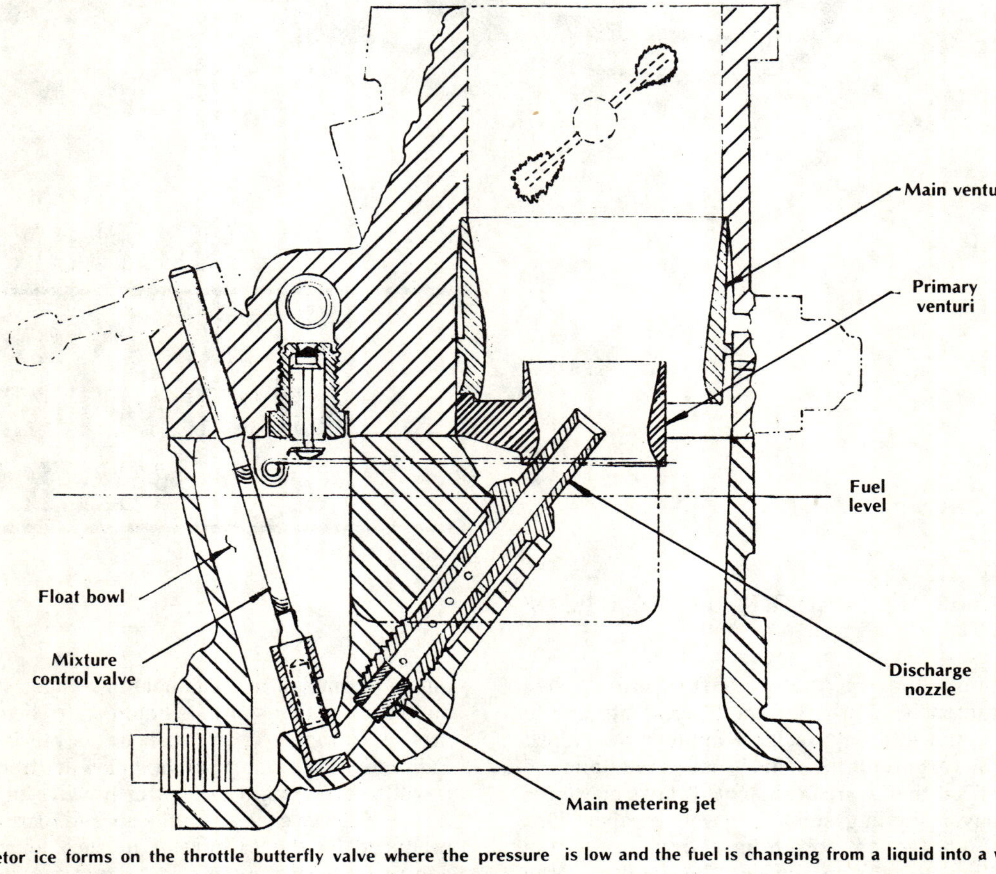

Carburetor ice forms on the throttle butterfly valve where the pressure is low and the fuel is changing from a liquid into a vapor.

- Figure 121 -

B. MIXTURE CONTROL BY EXHAUST GAS TEMPERATURE MONITORING:

"Lean the mixture until the engine gets rough, and enrich it until it smooths out," is an age old way of adjusting the carburetor mixture. This works well on engines where power output is relatively low, but with modern engines operated at power outputs reaching their design limits, this rule of thumb is neither efficient nor safe.

A basic fact of thermodynamics allows us to lean the mixture scientifically so we know exactly what is taking place in the cylinder. When fuel and air are mixed and the mixture burned, if there is 0.067 pounds of fuel for every pound of air, there will be the maximum amount of heat released. If the ratio of fuel to air is greater than this, less heat will be released because there is not enough oxygen to unite with all of the fuel. If the mixture is leaner than this optimum, the heat release will also be less because there is not enough fuel for the air.

A thermocouple probe similar to that used to measure the tail pipe temperature in turbine engines is stuck into the exhaust stacks just down stream from the exhaust valve. This measures the temperature of the gases coming out of the cylinder. The actual temperature is not the issue with this instrument, but the temperature of these gases relative to their maximum temperature. It is convenient that the maximum temperature always occurs with the same mixture ratio regardless of the percent of power used.

The procedure for adjusting the mixture for cruise is to climb to the desired altitude and throttle back to the power you want, normally somewhere between 55% and 70%. Then, adhering strictly to the engine manufacturer's recommendations, lean the mixture until the exhaust gas temperature peaks. When this happens, the mixture is 15:1 (0.067), and the heat energy release is maximum. Most manufacturers then recommend that the mixture be enriched to provide a temperature drop of about twenty-five to fifty degrees Fahrenheit. At lower powers, it may be allowable to operate at the peak EGT. The temperature of the exhaust valve is the limiting factor rather than the temperature of the gas itself. And the EGT must be decreased to maintain a constant valve temperature as the engine power is increased. Before operating at peak EGT, be sure the manufacturer of your engine allows it.

There are two types of exhaust gas temperature indicating systems, one using only one probe, and the other using a probe in each cylinder. The design of the induction system prevents any one cylinder being the leanest at all times, so with the multi-probe system, after the power reduction has been made, the cylinders are scanned to find the hottest one. Lean the mixture until the EGT of this hottest cylinder peaks, and then adjust the mixture according to the engine manufacturer's recommendations: enrich to get the desired drop from the extra fuel, leave it on peak, or lean to get a temperature drop because of air cooling.

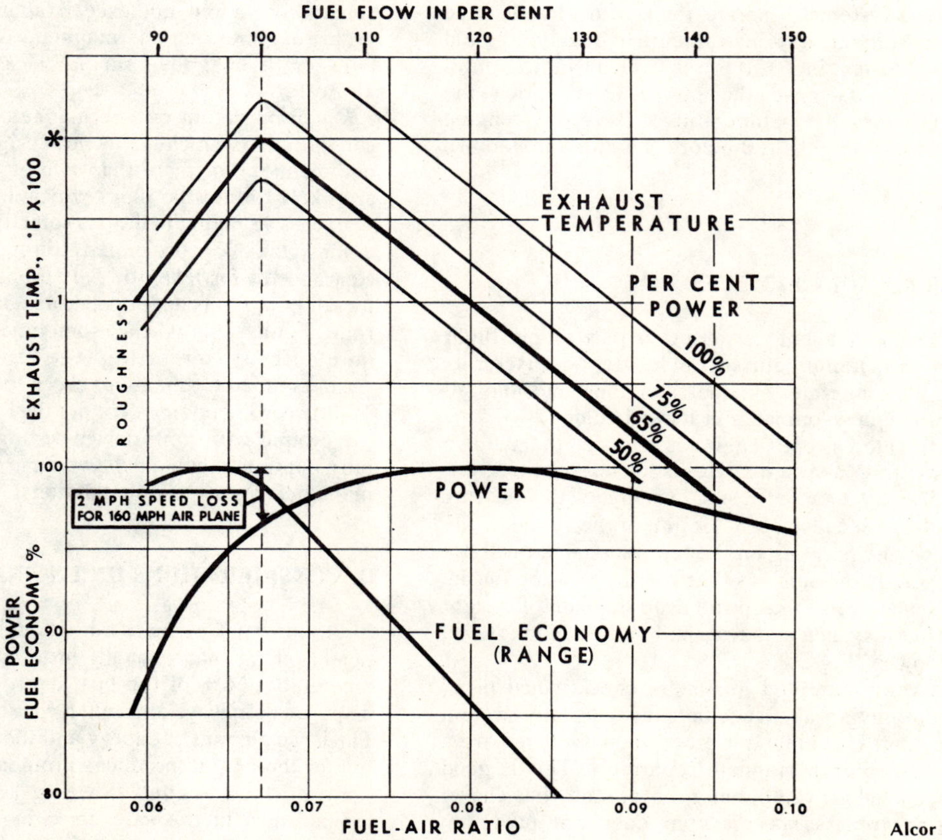

The peak exhaust gas temperature occurs with the most efficient fuel-air ratio, [15:1]. This creates more heat than most engines can tolerate at high power settings.

- Figure 122 -

Exhaust gas temperature gages normally do not have the temperature marked on the dial, but the temperature covered by the indicator is in the range of 1000 to 1800 degrees Fahrenheit. A calibration point, either a mark on the dial, or a movable pointer, is used to establish the maximum EGT. The instrument is calibrated by flying the airplane at an altitude where 65% power is developed at full throttle. After all of the temperatures have stabilized, lean the mixture until the EGT peaks. Adjust the indicator to the mark on the dial, or move the pointer to this temperature. This is now the maximum EGT that should ever be allowed. If the exhaust gas temperature is allowed to go higher than this, there is a danger of the exhaust valve becoming overheated to such a point that valve failure could occur.

The EGT system is not only useful for setting the correct mixture for cruise, but is an excellent instrument for trouble-shooting engine malfunctions which might arise in flight.

1. A rise in EGT with no change in operating conditions could indicate an ignition system malfunction. A rise in only one cylinder would indicate plug failure, or a rise in all cylinders could point to magneto failure.

2. A decrease in the EGT with no change in operating conditions could indicate carburetor ice. A plugged injector nozzle could give a decrease in EGT on one cylinder of an injected engine.

Since these systems generate their own electricity to drive the instrument, any instrument malfunction would cause a lowered reading. If there is ever an indication of excessive EGT, decrease the power, and enrich the mixture to lower the temperature. Have the engine checked by a competent A&P before the airplane is flown again.

C. PROPER MAGNETO CHECK:

A mag check is about as routine a part of pre-flight preparation as fastening our seat belt, yet, because of its familiarity, we sometimes overlook its importance and fail to adapt to new developments in its technology.

Each certificated aircraft engine is required to have two independent ignition systems. This is for safety in case one fails, but also for more efficient burning of the mixture in the cylinders. Because of the design of the cylinder, the power loss when operating on the top plugs alone will not always be exactly the same as that when operating on the lower plugs alone. So to keep the drop the same for right and left magnetos, each mag fires half of the top plugs and half of the lower plugs.

The generally accepted magneto check consists of operating the engine at a specific speed, usually around cruise rpm, and checking the drop when switched from both magnetos to each magneto separately. This is good and has been used for years, but a recent study has shown that checking from a specific rpm does not take into consideration the power variations brought about by changes in atmospheric conditions. For this reason, rather than checking the magnetos at a specific rpm, some engine manufacturers recommend that the magnetos be checked at a power setting of around 50 to 65% as indicated by both the tachometer and manifold pressure gage. Before using this system, be sure to consult your engine operator's manual for the proper rpm and manifold pressure.

A proper check is conducted by:

1. Starting the engine and allowing it to warm up as the manufacturer recommends. Low speed warm-ups are not usually recommended because of the tendency to foul spark plugs, and because the generator or alternator does not provide the needed electrical energy.

2. Be sure the propeller is in the full low pitch position so a drop in power will produce a drop in rpm rather than a change in propeller pitch.

3. Set the throttle to produce 50 to 65% power, or as specified by the engine manufacturer.

4. Turn the magneto switch from "Both" to "Left" and observe the drop in rpm; then put the switch back on "Both" until the rpm returns to its original value.

5. Move the switch from "Both" to "Right" and note the drop; then put it back on "Both."

The drop for each magneto should be within the limits specified by the manufacturer, and the engine should run smoothly while operating on each magneto. The difference between the rpm when operating on each magneto separately should not exceed about 50 rpm. Excessive drop, roughness on one magneto, or excessive difference between the magnetos should be called to the attention of an A&P.

When operating on one magneto, the engine is under conditions for which it was not designed. Don't leave it on one magneto for more than two or three seconds, or you are likely to foul the plugs which are not firing.

Excessive magneto drop is sometimes encountered after a long taxi, or prolonged idling while waiting for a clearance or for take-off. This problem can be minimized by idling at the proper speed, usually around 1200 to 1500 rpm. Fouled plugs can sometimes be burned out by momentarily operating the engine at the power recommended for magneto check, or perhaps a bit higher. Tightly cowled engines are not designed to be operated on the ground with high power settings, so any attempts to burn out fouled plugs must be done in accordance with the manufacturer's recommendations.

D. CONSIDERATIONS ON ENGINE COOLING:

An aircraft engine is a heat engine, and a highly inefficient one at that - usually less than 30% efficient. This means that 70% of the heat energy in the fuel must be disposed of in ways other than conversion into useful work. The large amount of energy and the light-weight construction of the engine produces problems in cooling.

Properly designed cowling and baffles keep the temperature in the safe operating range for almost all steady-state conditions, and careful monitoring of the cylinder head temperature and use of cowl flaps help control the steady temperatures, but it is the sudden temperature changes, thermal shock, that really does damage to an engine.

Scuffed pistons can result from a too-sudden application of power to a cold engine.

- Figure 123 -

If an engine is idled at low RPM for a long period of waiting for take-off, it will be relatively cold. When clearance is given, if the throttle is suddenly opened to full take-off power before any forward speed is gained, uneven expansion of the engine parts can cause the piston to scuff inside the cylinder or the exhaust valves to warp and stick inside the valve guides. If you must wait, keep the engine operating at the speed recommended by the engine manufacturer, and apply the power gradually, going to full throttle only after there is an airflow through the engine due to your forward speed.

In the climb-out following take-off, the airspeed should be kept as high as consistent with good operating practice, and the throttle left wide open until you are ready for your first power reduction. Almost all carburetors have provisions for the addition of extra fuel for cooling when the throttle is wide open. Backing off on the throttle just a little bit after lift-off can shut off this extra fuel and deprive the engine of the cooling it needs at this critical point.

It is not only bad practice to heat the engine too quickly, but cooling it too fast will also do damage. If a hot engine is throttled back and allowed to idle as the airplane glides down through cold air, the rapid change in temperature can cause cylinders and cylinder heads to crack. This is not only dangerous, but extremely costly. Enough power should be maintained during descent that the engine is not allowed to cool too rapidly, nor should the engine be allowed to become so cold that a power application during a go-around could produce a thermal shock.

QUESTIONS:

138. Is it possible for carburetor ice to form when the outside air temperature is well above freezing?
139. Under what condition is it proper to use partial carburetor heat when icing conditions exist?
140. What danger can arise from operating an engine at peak exhaust gas temperature at high power settings?
141. What would likely be indicated if the exhaust temperature on one cylinder suddenly increased when no operating controls were changed?
142. Why should an engine not be throttled back slightly as soon as the airplane breaks ground?

GLOSSARY

This glossary of terms is to give a ready reference to the meaning of some of the words with which you may not be familiar. These definitions may differ from those of standard dictionaries, but are more in line with shop usage.

A N hardware: Standard hardware items such as bolts, nuts, washers, etc., whose design and material have been approved by both the Air Force and Navy and are acceptable for use in civilian airplanes as well as military.

A & P Technician: An aircraft mechanic who has met the experience and knowledge requirements and is authorized by the FAA to return an airplane to service after certain inspections and maintenance procedures.

absolute pressure: Pressure referenced from zero pressure, or a vacuum. It is usually measured in inches of mercury.

accumulator: A hydraulic component consisting of two chambers separated by a piston, diaphragm, or bladder. Compressed air in one chamber holds pressure on hydraulic fluid in the other chamber. Hydraulic fluid under pressure can thus be stored in a system.

acrylic: A glossy, transparent thermoplastic used for cast or molded parts such as aircraft windshields.

actuator piston: The movable part of a hydraulic or pneumatic actuator.

aerodynamic shape: The shape of an object with reference to the airflow over it. Certain shapes cause air pressure differentials which produce lift; others are designed for minimum resistance to the flow of the air.

aileron: A control surface attached to the trailing edge of the wing which controls the airplane in a rolling movement.

airfoil: A surface which produces lift by the dynamic action of the air moving over the surface.

airframe: The airplane structure itself - fuselage, wings, tail unit, landing gear, and essential flight controls.

airspeed indicator: A differential pressure gage that measures the difference in ram air pressure and static pressure to provide an indication of airspeed in miles per hour or knots.

Airworthiness Certificate: A certificate issued by the FAA to all aircraft which have met the minimum standards required for certification.

Airworthiness Directive: A regulatory notice sent out by the FAA to the registered owner of an aircraft informing him of a condition which must be corrected for the aircraft to maintain its airworthiness status.

alternator: A special type of generator in which alternating current electricity is generated in the fixed windings and is converted to direct current by solid state rectifying diodes.

altimeter: An aneroid (no liquid) barometer calibrated in feet above a reference pressure. It measures the weight of the column of air above it.

aluminum alloy: Aluminum combined with other elements to produce a metal having strength superior to the original material.

anti-drag wire: A diagonal, load carrying member of a Pratt truss wing. It runs from the rear spar inboard to the front spar outboard and carries tensile loads tending to pull forward on the wing.

anti-servo tab: An adjustable tab attached to the trailing edge of a stabilator which moves in the same direction as the primary control. It is used to make the stabilator less sensitive.

approved repair station: A facility approved by the Federal Aviation Administration for certain types of repair to certificated aircraft.

Approved Type Certificate: An approval issued by the FAA for the design of an airplane, engine, or propeller. This certifies that the product has met at least the minimum design standards.

automatic direction finder [ADF]: An electronic navigation system which indicates to the pilot the bearing of a radio station relative to the nose of the airplane.

avionics: A word coined to mean airborne electronic equipment.

baffles: Air deflectors, between and around air-cooled cylinders to aid in uniform cooling.

battery bus: The electrical tie point in an airplane where power from the source is distributed to the various loads.

bayonet thermocouple probe: A pick-up for cylinder head temperature which presses into an adapter screwed into the side of a cylinder head.

bell crank: A double lever in a control system used to change the direction of motion. These are commonly used in an aileron system to change spanwise movement into chordwise movement to move the control surface.

bladder tank: A neoprene impregnated fabric bag attached inside of an airplane structure to hold fuel.

bogus parts: Parts not approved for use in a certificated airplane. These parts are often marked so as to appear to be authorized parts, but when they are installed in an airplane, safety may be compromised.

bourdon tube: A curved, thin wall metal tube in an instrument mechanism used to measure gage pressures. Pressure inside the tube tends to straighten it. This straightening action moves a pointer which indicates the amount of pressure.

brake caliper: The clamp in a disc brake system which holds the brake linings.

bungee shock cord: A shock absorbing medium composed of numerous rubber bands enclosed in loosely woven fabric.

calendar month: The measure of time used by the FAA for certification purposes. A calendar month ends at midnight of the last day of the month, regardless of the day it began.

Cannon Plug: A trade name for a quick-disconnect electrical plug containing a number of individual wires.

capacitance type fuel gaging system: A fuel quantity indicating system using the fuel as the dielectric of a capacitor (condenser). It actually measures the weight of the fuel instead of its volume.

catalyst: A material which is used to bring about a change, but which does not actually enter into the change itself.

center of gravity range: The allowable range in inches from the datum in which the center of gravity of an airplane is allowed to fall.

chevron seal: A single direction seal in a hydraulic or pneumatic actuator. It derives its name from its V-shaped cross section.

circuit breaker: A circuit protecting device which opens the circuit in the case of excess current flow. It differs from a fuse in that it can be reset without having to replace it.

commutator: The copper bars on the end of a generator armature to which the rotating coils are attached. Alternating current is generated in this armature, and brushes riding on the commutator act as a mechanical switch to convert it into direct current.

compensating port: A port inside a brake master cylinder which vents the wheel cylinder to the reservoir when the brake is not applied. It prevents fluid expansion due to heat causing the brakes to drag.

compression strut: A structural member - a short bar or tube, etc. - separating the spars in the wing and bearing the compressive loads.

compressive load: A load or a force which tends to compress or squeeze an object together.

corrosion: An electro-chemical process in which a metal is transformed into chemical compounds which are powdery and have little mechanical strength.

Dacron [Ceconite, Eonnex, Poly-Fiber]: The proprietary name of a synthetic textile fiber, noted for its great resilience.

datum: An arbitrary reference line from which all measurements are made when determining the moments used for weight and balance computations.

differential compression check: A test of the condition of an engine cylinder in which leakage past the piston rings and valves is determined by measuring the pressure drop across a calibrated orifice in the tester. The tester is in series with the engine cylinder.

diode: A solid state electron check valve which allows electron flow in one direction but stops it in the opposite direction. Six diodes are used to convert AC into DC in an aircraft alternator.

directional gyro: An attitude gyroscopic instrument which provides a directional reference for the pilot. It is not directionally sensitive, but must be set to agree with the magnetic compass. It does not have the oscillations or swings of a magnetic compass.

Distance Measuring Equipment: Electronic navigation equipment which measures the time required for a signal to travel from the airplane to a ground station and return to the airplane. This time is translated into nautical miles to the station.

dope: The finishing material used on fabric surfaces, serving to tauten and strengthen the fabric as well as making it weatherproof.

doubler: A piece of sheet metal placed against an aircraft skin where some component is attached, to provide stiffness or additional strength.

drag wire: A diagonal, load carrying member of a Pratt truss wing. It runs from the front spar inboard to the rear spar outboard and carries tensile loads tending to drag back on the wing.

earth's magnetic field: Magnetic lines of flux which surround the earth. These lines enter and leave the earth at the magnetic poles, different from, but near, the geographic poles.

elevator: The horizontal movable control surface in a conventional aircraft empennage. It is used to cause the airplane to rotate about its lateral axis, to pitch.

Emergency Locator Transmitter: A small, self-contained radio transmitter that will automatically, upon impact of a crash, transmit an emergency signal on 121.5 and 243.0 mhz.

empennage: The tail of an airplane.

empty weight: The weight of the airplane with all fixed

equipment installed. All necessary fluids except usable fuel and drainable oil are included.

empty weight center of gravity: The center of gravity of the airplane with all fixed equipment installed, but with only unusable fuel and undrainable oil on board.

FAA Form 337: Major Repair and Alteration Form.

fairleads: Wood or plastic guides for aircraft control cable used to hold the cable away from the structure in straight runs.

FAR s: Regulations issued by the Federal Aviation Administration. These are to be followed by all aircraft pilots, technicians, and manufacturers of civil aircraft. The purpose of the FAR s is to ensure maximum safety in the operation of all commercial and private aircraft.

FCC: Federal Communications Commission. That part of the federal government that regulates the transmission of energy or intelligence by radio.

Federal Aviation Administration [FAA]: That branch of the federal government which monitors and controls the certification of civilian airmen and aircraft.

flaps, Fowler: Wing flaps which are lowered by sliding out of the trailing edge of the wing on a track. They modify the shape of the airfoil and increase the area of the wing.

flaps, simple: Wing flaps which are lowered by pivoting them about a point near their leading edge. They change the airfoil section of the wing but do not affect the wing area.

flight manual: Approved information which must be carried in an airplane. This pertains to the speed and load limits and to any other information that is vital to the pilot.

formers: Frames of light wood or metal which attach to the truss of the fuselage or wing in order to provide the required aerodynamic shape.

"fresh" annual inspection: An annual inspection which has just been performed on an airplane. Sometimes this is used as a selling point for an airplane. It is actually meaningless unless you know the integrity of the person performing the inspection.

fuselage: The body of an airplane. That part to which the wing, tail, and landing gear attach, and which in a single engine airplane usually carries the engine.

generator: An electrical device that converts mechanical energy into electrical energy by electromagnetic induction.

gimbal: The frame in which a gyro wheel is mounted.

glide slope: The electronic portion of an Instrument Landing System (ILS) which provides the pilot with the proper descent angle for his approach.

Grade A cotton: The standard fabric used for covering truss type aircraft structure. It is made of long staple cotton, has 80 threads per inch, and a tensile strength of 80 pounds per inch.

gyro horizon: An attitude gyroscopic instrument that indicates rotation about the pitch and roll axes.

hydraulics: That branch of power transfer in which liquids under pressure are caused to produce work.

hydro-sorb: A hydraulic shock absorber used in a bungee shock cord landing gear to prevent rebound.

Inspection Authorization [IA]: An authorization issued by the FAA to experienced A&P mechanics meeting certain requirements. This allows them to return aircraft to service after annual inspections or certain major repairs.

Instrument Flight Rules [IFR]: Flight by reference to instruments rather than outside visibility. These rules apply when the ceiling and/or visibility are below certain minimums set by the FAA.

Instrument Landing System: The electronic navigation system which provides a pilot with a narrow indication of the centerline of the runway, a glide slope to descend, distance markers along the approach, and high intensity lights to aid in the transition to the visual portion of the approach.

integral fuel tank: A portion of the aircraft structure, usually a wing, which is sealed off and used as a fuel tank.

jackscrew: A screw controlled from the cockpit which raises or lowers the leading edge of an adjustable stabilizer.

laminated wood: Layers of wood glued together so that the grain in all layers runs in the same direction.

landing gear: The wheels, floats, skis, and all of the attachments which support the airplane when it is resting on the ground or water.

lateral axis: The pitch axis of an airplane; that axis which extends through the center of gravity of an airplane from wing tip to wing tip.

lead, ignition: A special, heavily insulated, shielded wire that connects the magneto to the spark plug.

leading edge: The front edge of an airfoil.

localizer: The electronic portion of an Instrument Landing System (ILS) that indicates the centerline of the runway to the pilot.

logbooks - mechanical: Books containing records of total operating time, repairs, alterations or inspections performed, and all A.D. Notes complied with. One should be kept for the airframe, each engine, propeller, and one for avionics and instruments.

longitudinal axis: That axis of an airplane which extends through the fuselage from nose to tail, passing through the center of gravity. It is also called the roll axis.

magneto: An AC generator and step-up transformer which produces a high voltage to the spark plug at a specific time relative to the piston position within the cylinder.

main bus: A common tie point for electrical circuits to obtain their voltage.

manifold pressure: The absolute pressure, measured in inches of mercury, existing in the intake manifold of an engine. This is the pressure that forces the fuel-air mixture into the cylinder.

marker beacon: An electronic navigation system that indicates a specific location by a highly directional signal, receivable only directly above the transmitter. This signal illuminates an indicator light and provides an audible tone.

modulated: Changed. This normally refers to a radio carrier being changed by the voice or by a tone.

moment: The product of the weight of an object in pounds, and the distance from the center of gravity of the object to the datum in inches. Moment is used in weight and balance computations and is expressed in pound-inches.

monocoque: A stressed skin type of construction in which the stiffness of the skin provides a large measure of the strength of the structure. No truss or substructure is required.

octane rating: An expression of the anti-detonation characteristics of aviation gasoline.

oleo strut: An aircraft landing gear shock strut which absorbs the initial landing impact by the transfer of oil from one chamber to another through a restricting orifice. Taxi shocks are absorbed with compressed air or by a spring.

Omni Bearing Selector [OBS]: The control with which a pilot can determine the VOR radial on which he is located.

omnirange navigation equipment [VOR]: A phase comparison type of electronic navigation equipment which provides a directional reference between the airplane and the ground station, measured from magnetic north.

Operations Limitations Manual: See **flight manual.**

O-ring: A round cross-sectioned seal used in hydraulic and pneumatic systems. O-rings seal in either direction.

Par-Al-Ketone: A waxy grease used to protect control cables and hardware fittings from corrosion on seaplanes.

parasite drag: Drag caused by the friction of the air flowing over a body. Parasite drag increases as airspeed increases.

plastic: A commonly used term to denote any of the thermoplastic or thermosetting polymers used in modern aircraft construction.

plasticizer: A constituent of aircraft dope used to give it resilience.

plywood: Layers of wood glued together so that the grain in each layer is placed 45° or 90° to the next.

polyurethane enamel: A hard, chemically resistant finish whose long flow-out time and even cure throughout give a flat surface and a glossy "wet" look. Especially suitable for seaplanes and agricultural planes because of its resistance to chemical action and abrasion.

Pratt truss: A type of truss structure in which the vertical members carry only compressive loads, and the diagonal members carry only tensile loads. A Pratt truss is used for most fabric covered wings.

precession: That characteristic of a spinning wheel such as a gyro that causes a force to be felt 90 degrees to the point of application in the direction of rotation. In the case of a directional gyro, bearing friction causes precession to drift the indicator.

priority valve: A pressure actuated hydraulic valve which allows certain actuations before others. This type of valve is used to assure that the wheel well doors will be opened before the landing gear extends.

prototype: The first unit of a design to be built.

puck: The brake lining of a Goodyear single-disc brake. A small round or kidney-shaped block of asbestos-type material.

Radar Beacon Transponder: A piece of electronic equipment which receives an interrogation from ground radar and responds with a coded transmission that appears on the radar scope of the traffic controller.

radome: A plastic enclosure for the radar antenna.

rate of climb indicator: A rate of pressure change indicator which indicates the rate in feet per minute. It is also called a vertical speed indicator.

rate of yaw: The rate, in degrees per second, the airplane rotates about its vertical axis.

Registration Certificate: The document in the airplane which contains the name and address of the person to whom the airplane is registered.

relief valve: A pressure actuated hydraulic or pneumatic valve which will open to relieve excessive pressure on the system.

retraction test: That portion of an aircraft inspection in which the airplane is put on jacks and the landing gear cycled through its retraction and extension sequence.

RPM: Revolutions per minute. Rotational speed of the engine crankshaft.

rudder: The movable vertical control surface. It is used to rotate the airplane about its vertical axis.

rust: Oxidation of iron. A red, crusty product which forms on iron or steel when it unites with oxygen.

selector valve: A hydraulic flow control valve which directs hydraulic pressure to one side of an actuator and connects the other side to the system return line.

semi-conductor diodes: See **diodes.**

semimonocoque: A stressed skin structure in which the skin is supported by a light-weight framework to provide extra rigidity. Most of the larger modern aircraft are constructed with semimonocoque structure.

sequence valve: A mechanically actuated hydraulic valve that causes a sequential action of certain actuators. Wheel well doors, for example, must open and contact a sequence valve before the landing gear can extend.

servo tab: An adjustable tab attached to the trailing edge of a control surface. The tab moves opposite the direction of the control. It is used to aid the pilot in moving the control.

snubber: That portion of a hydraulic actuator which arrests the motion of the piston at the end of its stroke to cushion the stopping action.

spar: The principle spanwise member of a truss type wing structure.

spar varnish: A phenolic modified oil which cures by oxidation rather than evaporation of its solvents, producing a tough, highly water-resistant film.

stabilator: A single piece horizontal tail plane which combines the functions of a stabilizer and an elevator.

stabilizer: An airfoil; usually the fixed horizontal tail surface.

static system: Plumbing that connects the altimeter, airspeed indicator, and vertical speed indicator to the outside static air source of the airplane. An alternate source is usually included in this system.

stator: The fixed windings of an alternator in which the load current is generated.

stiffener: A reinforcement attached to an airplane structure to increase its resistance to bending.

stressed skin structure: A structure in which the covering, the skin, is designed to carry the tension and compression stresses of structural loads.

stringers: Thin metal or wood strips running the length of the fuselage to fill in the shape of the formers. Longitudinal stiffeners.

superstructure: A framework attached to an aircraft truss structure to provide the desired aerodynamic shape. It is usually covered with lightweight sheet metal or aircraft fabric.

Supplemental Type Certificate [STC]: A certificate authorizing an alteration to an airframe, engine, or component which has been granted an Approved Type Certificate.

system pressure regulator: That hydraulic component which controls the hydraulic system pressure. It unloads the pump when a pre-selected pressure is reached and directs the pump back onto the line when the pressure drops to the desired kick-in pressure.

tab: An auxiliary airfoil hinged to a control surface, used to assist in maneuvering or trimming, etc.

tachometer: An instrument which measures the rotating speed of the engine in revolutions per minute (RPM).

take-off power: Brake horsepower approved by the engine manufacturer for take-off. This may be limited in the time allowed.

tensile load: An external force tending to lengthen or stretch a body.

tensiometer: An instrument used to determine the tension on aircraft control cables.

thermoplastic resin: A resin material which will soften with the application of heat. Most windshields and side windows are made of this material.

thermosetting resin: Most widely used in today's plastics, it sets usually by chemical means and maintains its hardness even when heat is applied.

torque: A force that produces or tends to produce rotation.

torque link: The hinged linkage between the piston and cylinder of an oleo shock strut. The piston is allowed to move in and out but is restrained from rotating. These are also called scissors or nutcrackers.

torsion: An external stress which produces twisting within a body.

trailing edge: The rear edge of an airfoil.

transceiver: A piece of electronic communications equipment in which the transmitter and receiver are housed in the same unit.

trim tab: A small hinged inset in the trailing edge of an aileron, elevator, or rudder; separately controlled, to balance or trim the airplane.

truss fuselage: A fuselage constructed usually of welded steel tubing to carry the tensile and compressive loads. A superstructure or auxiliary framework is often attached to the truss to give the structure a desirable aerodynamic shape.

turbocharger: An air compressor driven by exhaust gases which increases the density of the air going into the engine through the carburetor or fuel injection system.

turn and slip indicator: A rate gyro which senses rotation about the vertical or yaw axis. A glass ball in a curved glass tube indicates the relationship between the angle of bank and the rate of yaw.

turnbuckle: A device consisting of a barrel and two forks having opposite cut threads, inserted in a control cable to adjust its tension.

turn coordinator: A rate gyro which senses rotation about both the roll and yaw axes.

Type Certificate Data Sheets: The official specifications of an airplane, engine, or propeller. These are issued by the FAA, and the device, in order to be airworthy, must conform to these specifications.

unloading valve: See **system pressure regulator.**

vertical axis: The axis of an airplane extending vertically through the center of gravity. It is also called the yaw axis.

V tail surface: An empennage consisting of two fixed and two movable surfaces arranged in a V shape. These two surfaces have the same aerodynamic function as the more conventional three surfaces.

whiffletree: A steering bell crank.

wing flaps: Movable control surfaces on the trailing edge of a wing, inboard of the flaps, which alter the airfoil section of the wing, increasing the lift and the drag.

yaw: Rotation about the vertical axis of an airplane.

ANSWERS TO STUDY QUESTIONS

1. The registered owner or operator of the aircraft.
2. The engine Type Certificate Data Sheets.
3. Airworthiness Directives are sent to the registered owners of each affected aircraft, and a summary of all A.D. s is available for A&P s and I.A. s.
4. An FAA Form 337 must be made up for each major alteration of an aircraft. The original of this form is kept with the aircraft records.
5. A record of all A.D. Note compliances must be kept with the aircraft records. It must show the date and the method of compliance.
6. An A&P mechanic holding an Inspection Authorization.
7. The inspections are the same, the only difference being the person allowed to perform them. An A&P with an Inspection Authorization is required to perform the Annual inspection.
8. Only until the next Annual inspection is performed.
9. For one year.
10. Every 24 calendar months.
11. a. Registration Certificate.
    b. Airworthiness Certificate.
    c. Operations Limitations.
    d. Radio Station license.
12. If a flight manual is required for a particular airplane, it will be listed in the Type Certificate Data Sheets for that airplane.
13. A title search can be made at the Aircraft Registry Section of the FAA in Oklahoma City to determine the registered owner of the airplane and a record of any liens against the airplane.
14. The fact that an airplane has had a "fresh annual" is no assurance of its condition. Only the integrity of the person performing the inspection makes it meaningful.
15. The application for registration is mailed to the FAA Aircraft Registration Branch in Oklahoma City, Oklahoma, along with a copy of the bill of sale.
16. No, the aircraft radio station license is not transferable to the new owner of an airplane.
17. A truss type structure normally requires a superstructure to give it its desired aerodynamic shape.
18. The strength of a monocoque structure lies in its rigidity; therefore, dents can weaken it.
19. Aluminum alloy is susceptible to corrosion.
20. This is a sheet of aluminum alloy with a coating of pure aluminum rolled onto both of its sides.
21. Magnesium alloy is highly susceptible to corrosion and cracking.
22. Cadmium plating protects steel hardware from corrosion.
23. In laminated wood, the grain of each layer runs in the same direction, while in plywood the grain of alternate layers are oriented 45 or 90 degrees to each other.
24. A thermoplastic material will soften when heat is applied, but a thermosetting material, once it is formed, cannot be softened by heat.
25. Aluminum pigmented dope prevents ultraviolet rays of the sun damaging the clear dope and the fabric.
26. Water penetrates the film of zinc chromate primer and released chromate ions which inhibit the formation of corrosion.
27. Wash primer is used to prepare aluminum alloy surfaces for some of the newer finishes.
28. Enamels have their pigments suspended in a varnish base which cures by oxidation. Lacquer pigments are suspended in a nitrocellulose film base and cure by the evaporation of the solvents.
29. Compressed air.
30. The transfer of oil from one compartment in the strut to another, through a restricted orifice.
31. Aircraft hydraulic fluid.
32. An O-ring between the wheel halves seals against the loss of air.
33. The compensating port allows for the expansion of fluid and eliminates brake drag.
34. Yes, provided the retreading is done by an FAA approved repair station.
35. a. A hand crank or lever.
    b. Compressed air.
36. The aileron moving up.
37. In the same direction as the control surface.
38. Fowler flaps slide out of the wing on a track, whereas simple flaps merely pivot about a point.

39. Stabilizers with adjustable leading edges vary the tail load for trim adjustment.

40. A pump could pull air from an empty tank, causing the engine to stop because of fuel starvation.

41. A portion of the wing is sealed off and used as a fuel tank.

42. The fuel lines are pressurized by auxiliary pumps in the fuel tanks.

43. a. Engine driven.
    b. Electric motor driven.

44. In case the engine pump pumps all of its fluid overboard, there will be fluid for emergency gear extension by the hand pump.

45. The unloading valve relieves the pump of any load when the system is not being operated.

46. An accumulator stores hydraulic fluid under pressure.

47. Red.

48. In case the unloading valve malfunctions, a relief valve will prevent damage to the system.

49. A priority valve allows certain hydraulic components to act sequentially, such as wheel well doors opening before the gear extends.

50. The commutator and brushes change alternating current into direct current.

51. Generator field current is controlled to regulate the output voltage.

52. The solid state rectifying diodes.

53. a. The heat caused by the current flow.
    b. Voltage drop caused by the resistance of the wire.

54. Shielding prevents interference from electromagnetic radiation.

55. Dilute sulfuric acid.

56. The absolute pressure in the intake manifold of an aircraft engine.

57. The fuel pressure at the fuel injector nozzles.

58. A thermocouple probe inserted into the exhaust stack.

59. Airspeed indicator, altimeter, and vertical speed indicator.

60. The differential pressure between ram air and static air.

61. The rate of pressure change as the aircraft climbs or descends.

62. The turn and slip indicator senses rotation about the vertical (yaw) axis, while the turn coordinator senses rotation about both the vertical and longitudinal (roll) axis.

63. Local magnetic fields caused by wires carrying electrical current or by magnetized portions of the airplane.

64. The slight unbalance of the compass card used to compensate for the dip error.

65. The relationship between the nose of the airplane and the spinning gyro wheel.

66. The pitch and roll axes.

67. A horizontally oriented V dipole antenna.

68. The same antenna used for the VOR receiver.

69. An ultrahigh frequency dipole antenna.

70. The glide slope receiver is automatically tuned when the pilot tunes the localizer.

71. 75 megahertz.

72. A small canoe-shaped antenna tuned to 75 megahertz.

73. A long wire sense antenna and a loop antenna.

74. A fixed loop or a rotating loop.

75. A UHF stub or blade antenna.

76. On the bottom of the aircraft fuselage.

77. A small, flexible wire whip antenna.

78. About twenty or thirty degrees of crankshaft rotation before the piston reaches top center.

79. Cast aluminum alloy.

80. An engine cylinder whose walls are case hardened by infusing ammonia into the surface of the metal.

81. Cast iron.

82. To aid in transferring heat from the valve head into the guide and then into the cylinder head.

83. To minimize valve train noise and wear.

84. To direct cooling air through the fins to remove the maximum amount of heat.

85. By keeping all of the contaminants in the oil in suspension, not allowing them to stick to the engine parts.

86. To provide for more uniform burning of the fuel air mixture in the cylinders, minimizing the danger of detonation.

87. The breaker points open, causing a rapid collapse of magnetic flux. This induces a high voltage into the secondary windings which causes a spark to jump the spark plug gap.

88. The magneto is "Off," or grounded.

89. Fine wire plugs are less affected by lead fouling from the tetraethyl lead in the fuel.

90. It minimizes the secondary current flow, decreasing the spark plug electrode erosion.

91. The phenolic or ceramic terminal of an ignition wire in a shielded spark plug.

92. A heavy clock-like spring spins the magneto after its normal firing position to provide a hot, late spark for starting an engine.

93. One set of points is in the starting circuit to provide a retarded spark for starting.

94. It increases the critical pressure and temperature of the fuel, allowing the use of higher cylinder pressure, thus more power from an engine.

95. Low lead 100 octane aviation gasoline contains a maximum of two milliliters of lead per gallon, as compared with 4.6 milliliters per gallon allowed for regular 100 octane fuel.

96. Since a carburetor meters its fuel according to the volume of air that flows into the engine, and the engine burns the fuel according to a weight ratio, fewer pounds of air mix with the fuel as the air becomes less dense; the mixture therefore becomes richer.

97. The additional fuel added to the mixture for full power removes heat from the cylinder heads.

98. A major advantage of a fuel injection system over a carburetor is its more even fuel-air mixture distribution to the cylinders.

99. An air metering force which is the difference between ram air pressure and the pressure drop in the venturi. This is a measure of the mass of air entering the engine.

100. The RPM of the engine.

101. The heat from the oil aids the vaporization of the fuel and minimizes the danger of carburetor ice.

102. A small exhaust driven turbine.

103. The propeller pitch.

104. Counterweights tend to move the propeller blades into high pitch.

105. No. The FAA does not give any mechanic the right to supervise an inspection. He must do the inspection himself.

106. Yes, a check list is required.

107. A compression check.

108. Small shavings on a freshly overhauled engine are caused by the parts fitting themselves together and are normally not a cause for concern.

109. a. Overheated spark plugs warn of an induction leak.
b. Oily plugs indicate worn valve guides or broken piston rings.
c. Sooty spark plugs could indicate operation with an overly rich mixture.
d. Excessive lead deposits are caused by the wrong heat range of spark plug.

110. High voltage can travel along a fingerprint on a spark plug cigarette and cause the plug to misfire.

111. By controlling the generator or alternator field current.

112. A local hot-spot which could be caused by detonation or a poorly fitting cylinder baffle.

113. They are usually replaced when found cracked.

114. The FAA issues General Aviation Inspection Aids which point out potential trouble areas.

115. Replace the belt if it is questionable.

116. It appears as a white powder on the surface or as puffiness under the paint film.

117. This indicates the bolt has been torqued and inspected.

118. Only an FAA approved repair station, one approved for this particular operation.

119. Every 24 calendar months.

120. No lubricant should be used between the bead and rim when installing a tube type tire.

121. MIL-H-5606 aircraft hydraulic fluid.

122. No. Nitrate dope will not stick to butyrate dope.

123. When selecting a primer, use only the primer recommended by the manufacturer of the paint being used. Follow their recommendations in detail.

124. Tighten the axle nut enough to remove all of the play from the bearing, back it off one castellation, and cotter key it.

125. MIL-H-5606 aircraft hydraulic fluid.

126. TSO-C-22.

127. By three colored threads woven into the braid which encases the rubber bands.

128. Yes, a seat is a structural part of an airplane.

129. A small, clip-over type spring steel nut called a Tinnerman nut.

130. If you should find metal particles in your oil screen, be sure to consult a reputable A&P mechanic. It may be an indication of real engine trouble.

131. This could indicate an induction system leak.

132. The lead deposits should be removed with a sharp metal probe or with a vibrator type spark plug cleaner; then the plug may be **lightly** abrasive blasted.

133. These tiny cracks are called "crazing" and occur in a curved plastic surface which has been exposed to extremes of temperature.

134. Any water sufficiently pure for drinking is usually suitable for use in an aircraft battery.

135. By the specific gravity of the electrolyte.

136. Always remove the ground cable first and install it last.

137. Safety of flight is compromised when a bogus part is installed in an airplane, not when it is purchased.

138. Yes, since the carburetor acts as a mechanical refrigerator it is possible for carburetor ice to form when the outside air is well above freezing.

139. Only when a carburetor mixture temperature indicator is installed.

140. Operating an engine at peak exhaust gas temperature can result in damage to the exhaust valves.

141. This could be an indication of a spark plug failure.

142. When the throttle is brought back, the extra fuel metered for cooling is cut off. The engine could overheat in a prolonged climb.

FINAL EXAMINATION

STUDENT ______________________________

GRADE ______________________________

Place a circle around the letter for the answer which is correct in each of the following questions.

1. What is true regarding the 100 hour inspection and an annual inspection?

   A. An annual inspection is more in-depth than a 100 hour inspection.
   B. These inspections are the same, the only difference being in the one authorized to perform them.
   C. Annual inspections are not required for aircraft operated for hire; only 100 hour inspections are.
   D. A check-list need not be followed when performing a 100 hour inspection if the mechanic is familiar enough with the airplane.

2. Air leakage through a wheel with a tubeless tire is prevented by:

   A. Using only one piece wheels for tubeless tires.
   B. An O-ring is installed between the wheel halves.
   C. Sloshing the inside of the tire and the wheel with a sealing compound.

3. What is an integral fuel tank?

   A. A tank in which the fuel is contained in a rubber bladder or liner.
   B. A portion of the wing which is sealed off around each rivet and seam and used as a fuel tank.
   C. A welded-up aluminum tank held in the structure by steel straps.
   D. A fiberglass tank attached to the wing tip.

4. What is the purpose of an accumulator in the hydraulic system?

   A. It stores hydraulic fluid under pressure.
   B. It generates hydraulic pressure.
   C. It prevents an overload on the pump.

5. Which statement is **NOT** true regarding generators and alternators?

   A. An alternator produces AC and a generator produces DC.
   B. The load current is produced in the rotating windings of a generator.
   C. The load current is produced in the stationary windings of an alternator.
   D. A generator uses a form of mechanical switch to change AC into DC.

6. Which instruments are connected to the aircraft static system?

   A. Altimeter, airspeed indicator, and turn and slip indicator.
   B. Directional gyro, gyro horizon, and turn coordinator.
   C. Suction gage, manifold pressure gage, and altimeter.
   D. Airspeed indicator, altimeter, and rate of climb indicator.

7. Which statement is true regarding the turn and slip indicator and the turn coordinator?

   A. The turn and slip indicator senses rotation about the roll axis only.
   B. The turn and slip indicator senses rotation about both the roll and yaw axis.
   C. The turn coordinator senses rotation about the yaw axis only.
   D. The turn coordinator senses rotation about both yaw and roll axes.

8. The same antenna is used for which two navigation systems?

   A. Omni and ADF.
   B. DME and marker beacon.
   C. Omni and ILS localizer.
   D. ILS localizer and ILS glide slope.

9. Why do some aircraft engines use hollow, sodium filled exhaust valves?

   A. They are stronger than solid steel valves.
   B. They transfer heat into the cylinder heads better than solid valves.
   C. They weigh less than solid steel valves.
   D. They hold the heat better than solid steel valves.

10. Why do aircraft engines use two spark plugs in each cylinder?

    A. Only for safety in case one ignition system fails.
    B. Only to satisfy a requirement of the Federal Aviation Administration.
    C. To prevent detonation by a more even burning of the mixture within the cylinder.
    D. To allow the engine to burn a fuel with a higher octane rating.

11. What is true about low-lead 100 octane aviation gasoline?

A. It can be used interchangeably with either regular 100 octane fuel or 80 octane aviation gasoline with no limitations or restrictions.
B. Low-lead 100 octane aviation gasoline is colored purple.
C. Low-lead 100 octane gasoline has less lead than 80 octane gasoline.
D. Low-lead 100 octane gasoline has more lead than 80 octane gasoline, but considerably less than conventional 100 octane gasoline.

12. Why is additional fuel fed into an engine when full power is being developed?

A. The additional fuel is used to get the extra power.
B. The additional fuel is used for cooling under full power operation.
C. The additional fuel helps the mixture burn faster than normal.
D. The additional fuel prevents the spark plugs fouling during full power operation.

13. What is true about counterweight type propellers?

A. Counterweights tend to move the blades toward high pitch.
B. Counterweights tend to move the blades toward low pitch.
C. Counterweights cannot be used on feathering propellers.
D. Only propellers having counterweights can be feathered.

14. Under what conditions may a private pilot legally perform an annual inspection on his own airplane?

A. Under the supervision of a licensed A&P mechanic.
B. Under the supervision of an A&P mechanic with an Inspection Authorization.
C. This is normally considered preventive maintenance, and he can do it as long as the airplane is not operated for hire.
D. Under no conditions, unless he also holds an A&P license with an Inspection Authorization.

15. What could be indicated if the center electrode insulators of the spark plugs in one cylinder are much cleaner and whiter than those in the other cylinders?

A. The cylinder from which these plugs were removed is probably getting more fuel than the other cylinders.
B. This is an indication of lead fouling on these plugs.
C. A valve guide on this cylinder is probably worn, admitting oil into the combustion chamber.
D. This cylinder is running too hot, possibly from an air leak in the induction system.

16. How does corrosion look as it first appears on an airplane structure?

A. It usually starts as pits in the center of a skin panel.
B. It first appears as a white powder along the edges of the sheets.
C. It appears as a red scale on aluminum surfaces.
D. Corrosion always forms under lap joints, and it cannot be detected until the sheet is damaged beyond salvaging.

17. What type lubricant is used when mounting a tube type tire on an aircraft wheel?

A. No lubricant is allowed.
B. Liquid soap.
C. Clean engine oil.
D. Silicone grease.

18. What should be done if you find small metal particles in the oil screen of your engine?

A. Ignore them, as it is normal for engines to shed small particles of metal during their operation.
B. Ignore them unless a magnet shows them to be steel particles.
C. Check with a reputable A&P mechanic to determine the seriousness of this metal.
D. Any small metal particles in the oil strainer require that the engine be completely disassembled to determine their source.

19. How are lead deposits best cleaned from the firing end cavity of an aircraft spark plug?

A. Don't attempt to remove the lead deposits; it does not hinder the plug operation.
B. Clean it out with an abrasive blast using an aluminum oxide abrasive.
C. Clean it out with a small wire bristle brush.
D. Pick it out with a small steel probe, or use a vibrator type spark plug cleaning machine.

20. Which procedure is correct when removing a battery from an airplane?

A. Remove the ground lead before removing the "hot" lead.
B. Remove the "hot" lead before removing the ground lead.
C. It makes no difference which lead you remove first, as long as the master switch is "off."

ANSWERS TO FINAL EXAMINATION

1. B
2. B
3. B
4. A
5. A
6. D
7. D
8. C
9. B
10. C
11. D
12. B
13. A
14. D
15. D
16. B
17. A
18. C
19. D
20. A